NOMINALISTIC SYSTEMS

SYNTHESE LIBRARY

MONOGRAPHS ON EPISTEMOLOGY,

LOGIC, METHODOLOGY, PHILOSOPHY OF SCIENCE,

SOCIOLOGY OF SCIENCE AND OF KNOWLEDGE,

AND ON THE MATHEMATICAL METHODS OF

SOCIAL AND BEHAVIORAL SCIENCES

Editors:

DONALD DAVIDSON, *Princeton University*

JAAKKO HINTIKKA, *University of Helsinki and Stanford University*

GABRIËL NUCHELMANS, *University of Leyden*

WESLEY C. SALMON, *Indiana University*

HUMANITIES PRESS / NEW YORK

ROLF A. EBERLE
University of Rochester

NOMINALISTIC SYSTEMS

D. REIDEL PUBLISHING COMPANY / DORDRECHT-HOLLAND

SOLE DISTRIBUTORS FOR U.S.A. AND CANADA
HUMANITIES PRESS / NEW YORK

Library of Congress Catalog Card Number 78-131265

SBN 90 277 0161 X

All Rights Reserved
Copyright © 1970 by D. Reidel Publishing Company, Dordrecht, Holland
No part of this book may be reproduced in any form, by print, photoprint, microfilm,
or any other means, without written permission from the publisher

Printed in The Netherlands by D. Reidel, Dordrecht

ACKNOWLEDGEMENTS

The present work is the result of a complete revision and considerable expansion of my doctoral dissertation (bearing the same title) which was deposited at the University of California in Los Angeles in August 1965.

Special gratitude is due to Professor Donald Kalish for supervising the original dissertation and for offering valuable suggestions and encouragement in uncounted discussions. Professor David Kaplan has been of great help in clarifying the intuitions and in aiding the transition to a formal treatment. I am indebted to Professor Robert Yost for his critical examination of the traditional problem of universals which has sparked and guided my interest in the subject. My gratitude extends to Professor Alfred Horn for valuable suggestions concerning algebraic aspects of the problem discussed. I am also greatly indebted to Professor Richard Montague for teaching methods and a style of philosophizing which I continue to regard as exemplary.

A fellowship, awarded to me by the University of Rochester during the summer of 1969, has aided in the preparation of this book. My colleagues at the University of Rochester deserve lasting gratitude for their encouragement and stimulating discussions.

The shortcomings, which will undoubtedly be found in these pages, are my own responsibility.

TABLE OF CONTENTS

ACKNOWLEDGEMENTS	V
CHAPTER 1 / INTRODUCTION	1
1.1. Program	1
1.2. Historical Background	3
1.3. Formal Preliminaries	10
1.3.1. Summary of Elementary Set-Theoretic Notions	11
1.3.2. Formal Systems, their Syntax and Semantics	17
CHAPTER 2 / INDIVIDUALS	24
2.1. Outline of Goodman's Conception	24
2.2. Part-Whole Relations	31
2.3. Atomistic Universes of Individuals	36
2.4. The Leonard-Goodman Calculus of Individuals LGCI	44
2.5. The General Calculus CII of Atomistic Individuals	50
2.5.1. The Vocabulary and the Axioms of CII	50
2.5.2. Some Theorems of CII	52
2.5.3. The Semantics of CII	56
2.5.4. The Semantical Adequacy of CII*	58
2.6. Finite Sums and Products: The Calculus CIII	62
2.6.1. The Vocabulary and the Axioms of CIII	63
2.6.2. The Semantics of CIII	64
2.6.3. The Semantical Adequacy of CIII*	65
2.7. Infinite Sums and Products: The Calculus CIIII	67
2.7.1. The Vocabulary and the Axioms of CIIII	68
2.7.2. Some Theorems of CIIII	69
2.7.3. The Semantics of CIIII	70
2.7.4. The Semantical Adequacy of CIIII*	70
2.8. Non-Atomistic Universes of Individuals	73

2.9. The Non-Atomistic Calculus CIIV and some Extensions	77
2.9.1. The Vocabulary and the Axioms of CIIV	78
2.9.2. The Semantics of CIIV	80
2.9.3. The Semantical Adequacy of CIIV*	80
2.9.4. Extensions of CIIV	81
2.10. Sequential Individuals and the Calculus of Individual Relations	81
2.10.1. Order Implied by Different Part-Whole Relations	84
2.10.2. Order Implied by the Existence of Certain Composites	85
2.10.3. Order Implied by Positional Parts	89
2.10.4. Order Implied by Relational Individuals and their Calculus	92
CHAPTER 3 / ONTOLOGICAL COMMITMENT AND DESIGNATA OF EXPRESSIONS	**102**
3.1. Ontological Implications of Theories	102
3.2. One-Place Predicates and their Extensions	122
3.3. Truth Conditions and Relations of Predication	125
3.4. Expressions of Relations and Operations	135
3.5. Denotationless Symbols and Nominalistic Models	142
CHAPTER 4 / BUNDLES OF QUALITIES, QUALITIES, AND CONCRETA	**148**
4.1. Bundles of Qualities and their Calculus CB	148
4.1.1. Background	148
4.1.2. The Vocabulary of CB*	150
4.1.3. The Semantics of CB with Informal Discussion	151
4.1.4. The Axioms of CB	158
4.1.5. The Semantical Adequacy of CB*	160
4.2. Resemblance and its Calculus CR	166
4.2.1. Background	166
4.2.2. The Vocabulary and the Axioms of CR	168
4.2.3. The Semantics of CR	171
4.2.4. The Semantical Adequacy of CR*	172
4.3. Qualities and their Calculus CQ	175
4.3.1. Background	175

4.3.2. The Meta-Theory of CQ		177
4.3.3. The Interpreted System CQ*		181
4.4. Concreta		184
4.5. Further Problems and Conclusion		197
BIBLIOGRAPHY		200
SYMBOLIC NOTATION		204
INDEX OF NAMES		207
INDEX OF SUBJECTS		209

CHAPTER 1

INTRODUCTION

1.1. PROGRAM

It will be our aim to reconstruct, with precision, certain views which have been traditionally associated with nominalism and to investigate problems arising from these views in the construction of interpreted formal systems. Several such systems are developed in accordance with the demand that the sentences of a system which is acceptable to a nominalist must not imply the existence of any entities other than individuals.

Emphasis will be placed on the constructionist method of philosophical analysis. To follow this method is to introduce the central notions of the subject-matter to be investigated into a system governed by exact rules. For example, the constructionist method of investigating the properties of geometric figures may consist in formulating a system of postulates and definitions which, together with the apparatus of formal logic, generates all necessary truths concerning geometric figures. Similarly, a constructionist analysis of the notion of an individual may take the form of an axiomatic theory whose provable assertions are just those which seem essential to the role played by the concept of an individual in systematic contexts. Such axiomatic theories gain in interest if they are supplemented by precise semantical rules specifying the denotation of all terms and the truth conditions of all sentences of the theory. The relevance of formal developments is pointed out by extra-systematic discussions designed to show how notions which are formally introduced correspond to pre-systematic concepts and how problems analogous to ones raised in informal contexts can be solved within the system. This method of philosophizing will be employed throughout.

Although an extensive use of formal logic is essential to our approach, we realize that the topics to be discussed have been of interest primarily to philosophers rather than to pure logicians. For this reason, we shall try to use only a minimum of symbolic notation, to supply informal readings for all important symbolic statements, and to surround the formal developments with a continuous philosophical discussion. Thus, substantial portions of the book should become accessible even to readers who shun formalism by the simple expedient of disregarding all technical details. In addition, the most technical sections are marked with an asterisk and may be omitted.

After some historical and formal preliminaries, an explication will be attempted of Goodman's notion of an individual. The necessary truths peculiar to this notion receive axiomatic treatment in formal systems (calculi of individuals), whose semantics is developed purely within set theory. As a test of adequacy, the semantical soundness and completeness of these systems is demonstrated. Apparently for the first time, the general notion of a 'generating' or 'part-whole' relation is clarified, and both atomistic and non-atomistic systems of individuals relative to arbitrary generating relations are formalized. Novelty is also claimed for our treatment of individuals which differ with respect to the arrangement among their parts.

Nominalists, it is frequently held, cannot endorse theories which imply the existence of items other than individuals. In order to understand this position, criteria are examined which govern the ontological implications of theories. A criterion of this sort is proposed which seems to have advantages over previous formulations.

'Universals', in one sense, are taken to be non-individual designata of predicates. The nominalistic opposition to 'universals', in that sense, gives rise to a discussion concerning methods acceptable to a nominalist for interpreting first-order predicate calculi with schematic predicates. On the basis of this discussion, the notion of a 'nominalistic model' is introduced which satisfies the requirement that all extensions assigned to non-logical symbols are individuals. Allowance is made, in a semantics employing such models, for simple and compound denotationless terms and for predicates which fail to designate. The importance, to a nominalist, of such notions as 'matching', 'resembling', and 'resembling in a given respect' for the purpose of specifying truth conditions is pointed out. Some of these notions are formally interpreted, axiomatized, and used in the framework of a nominalistic semantics.

Certain restrictions on the semantics of formal systems are examined which go beyond the minimal demands of a Goodman-type nominalism, and interpreted systems are proposed which meet such additional requirements. Some of these systems are axiomatized and their completeness demonstrated. Among them, we mention especially a theory which is interpreted in accord with the view that all empirical things are bundles of qualities.

Our entire account of nominalistic systems will be given in the framework of a theory which is not itself nominalistic. We provide, so to speak, a description of nominalism in a Platonistic language. As this procedure suggests, nominalism will be investigated, but neither advocated nor opposed. Indeed, we shall treat realism and nominalism not so much as ontological doctrines, but rather as alternative proposals concerning the construction of theories whose relative merits may depend upon the purposes which the theories in question are intended to serve.

1.2. HISTORICAL BACKGROUND

Throughout the subsequent pages, our discussions will be analytic rather than historical. Readers who are interested in the history regarding the problem of universals are urged to consult the enormous literature devoted to that subject. Nevertheless, many topics to which we shall address ourselves seem to have a close affinity to historically given positions. For this reason, a brief recapitulation of some of these positions may lend plausibility to the claim that our own formulations are akin to certain traditional doctrines.

Awareness of some problems concerning universals can be traced back at least to Plato's doctrine of ideas. It is an instance of Plato's theory that particular things are great, or human, or beautiful, only if they partake of the ideas or forms of greatness, of humanity, or of beauty. Such ideas are distinct from the particular things which participate in them: ideas are non-physical; if they are in space and time at all, then they can be wholly in two places at once; they are not sensed, not subject to change, and also not mere mental entities. Classification of things is warranted by virtue of their participation in forms; and knowledge and conception have forms as their objects. In addition, forms are also ideals whose perfection can only be approximated by individuals. Whereas particulars 'participate in' or 'copy' ideas, the ideas, among one another, 'blend', are subject to 'division', and admit of hierarchical arrangements. Plato was aware of problems regarding the nature of these relations.

Aristotle defined a 'universal' as 'that which is of such a nature as to be predicated of many subjects' and an 'individual' (or 'substratum') as 'that which is not thus predicated'. Chief among the universals rank qualities and relations (which are predicated of subjects like Socrates and Plato), secondary qualities (genera and species which are predicated of, or include, classes), and essences. Unlike Platonic forms, universals are not ideals of perfection, exist only in so far as there are objects which exemplify them, and come to be known by induction from the experience of particulars.

The medieval controversy arose out of Porphyry's commentary on Aristotle. Porphyry, with regard to Aristotle's 'secondary substances' raised, but left unanswered, the following questions: (1) whether genera and species subsist outside the mind, or exist only in the mind, (2) whether, assuming that they subsist outside the mind, they are corporeal or incorporeal, and (3) whether, under the previous assumption, they are separated from sensible things or exist in sensible things.

With respect to their answers to Porphyry's questions, three medieval schools of thought are usually distinguished: *Medieval Realists* (such as John Scotus, Anselm, and William of Champeaux) held that universals are real entities subsisting independent of mind and sense. *Medieval Conceptualists* (most historians count Abelard among them) maintained that universals are real but mind-dependent entities whose existence in the understanding may, however, be due to similarities among objects. *Medieval Nominalists* (among which one counts Roscelin and Ockham) advocated that only particular things exist. Thus, members of a species were thought to have nothing in common, except possibly the name of the species which refers to each individual of the species.

The questions regarding the status of universals were raised in many different contexts. One of them concerns the signification of predicates and common names. Whereas the proper names, like 'Socrates', name concrete individuals, what, if anything, could be signified by such words as 'man' or 'human'? Realists replied that there is, in addition to each human being, some mind-independent species or kind, of which the words 'man' or 'human' are names. Conceptualists favored the reply that 'man' or 'human' designate the result of some mental abstraction. And nominalists usually held that such general words designate severally each particular human being.

In another context, the justification of classificatory statements was at issue. Realists found the justification satisfactory that various objects are truly called 'white' because there is some quality 'white' whose presence in each of the objects explains why the object is white. Conceptualists have felt that classifications are conceptual devices which are imposed on things rather than discovered in the extra-mental world; and the justification for imposing a classificatory scheme on things lies in its aid to understanding. Nominalists have commonly held that classificatory statements are justified if all objects which fall under a classification bear certain resemblance relations to one another.

From a different point of view, a problem of universals has been raised by inquiring into the nature of concepts, or objects of conception or knowledge. If I conceive of a unicorn, is there something of which I conceive? Is it the same thing of which you conceive? If a unicorn were to exist, would it resemble the object of my conception, or be an instance of it? Realists have usually held, and nominalists denied, that concepts or objects of conception are entities of such a sort that several persons can grasp or conceive of the *same* entity.

Not only was the presumed function of universals quite different in these various contexts, but the very distinction between universals and individuals was drawn in many ways. Of special interest to contemporary nominalism are attempts to characterize individuals in terms of certain identity conditions, so-called *principles of individuation*. It was essential to Aristotle's conception that individuals are 'numerically one'; and individuals were said to be numerically one either if they consist of the same matter, or if they are continuous and have unity of form. Thus, a whole A composed of the parts of a shoe would be identical with a composite B either (1) if B consists of the same parts of the shoe, or (2) if A is continuous with B, and both A and B have the form of shoes. Species, unlike individuals, were the same if their elements satisfy the same definition. St. Thomas Aquinas, emphasizing one of Aristotle's principles, maintained that concrete individuals are identical if composed of the same matter. Other philosophers claimed that individuals are rather determined by their matter and form jointly, or by their form alone.

It is doubtful whether the problems concerning universals, as they arose in classical or medieval times, have ever been formulated clearly and precisely. Aristotle's grammatical terminology of 'subject' and 'predication' has encouraged subsequently a widespread failure to distinguish use and mention. Universals were said to 'characterize', to be 'predicable of', to be 'in', or 'to apply to' individuals. Universal entities, general names and abstract ideas were often ill distinguished, if distinguished at all. The semantical issues concerning the reference of general terms were interlaced with metaphysical discussions about substance, matter and form or about 'the one and the many'. Discussions were entangled with theological questions, epistemological problems, and *a priori* psychology. Thus, a contemporary treatment of the problem, even if it is intended to reconstruct some earlier positions, cannot be expected to achieve both precision in its exposition and also complete historical accuracy.

Historians investigating modern philosophy tend to agree that realism prevailed among Continental Rationalists, while conceptualism or nominalism was advocated by British philosophers. Hobbes held that only words (predicates and common nouns) can be said to be universal, due to the fact that they apply to many particulars. Locke regarded universals as creatures of our own making, ideas capable of signifying or representing many particulars. Berkeley and Hume rejected not only non-mental universals, but even abstract ideas. According to Berkeley, a particular idea comes to serve as a 'general' one by being made to represent all other particular ideas of the same sort. Thus, in a geometric demonstration, a particular triangle becomes representative of all triangles, and interchangeable with them for the purpose of the demonstration, by our refraining from mentioning those of its properties which it does not share with other triangles. In contrast to this interesting and rather clear notion concerning the function of general terms, Hume held that a particular idea becomes 'general' due to our habit of annexing a certain term with many similar stimuli. In spite of these nominalistic tendencies, the British Empiricists endorsed sensible qualities which may be regarded as repeatable and abstract in that they could be constituents of, or common to, various sensible objects.

Nominalists, both in medieval and modern times, have favored parsimony with respect to the categories of entities which were admitted. Instead of countenancing individuals, and also properties, relations, and classes of individuals, they 'shaved off' all but individuals. Even then, preference might be given to concrete, material individuals or, in the case of British Empiricists, to combinations of qualities which are given to sense (among them especially occurrent, rather than dispositional ones). According to their temper, nominalists have tended to rule out unwanted categories of entities on the grounds that positing their existence is contradictory, non-sensical, devoid of explanatory power, lacking in simplicity, unverifiable, or just plain suspicious. Some of the arguments which have been marshalled against universals seem similar to ones which might be advanced against the hypotheses that there are ghosts, or an ether.

In recent times, the positions of realism and nominalism have received new formulations whose indebtedness to tradition is frequently not made explicit, and often not obvious. We mention here only very briefly a few contemporary trends toward nominalism, some of which will be discussed more fully at appropriate stages in our own development.

Recent nominalism has found a formally trained advocate in the Polish logician Stanislaw Leśniewski. Leśniewski rejected 'general objects' in the sense of objects which have all and only those properties which are common to several individuals. He refused to countenance an empty set, unit sets which differ from their elements, and generally any set of individuals which is not itself an individual. He constructed the systems of Protothetic, Ontology, and Mereology to give formal treatment to logic, to the copula 'is', and to relations between parts and wholes. These systems are strongly extensional, and their formulas are intended to be individual expressions which cannot recur in different locations. Portions of mathematics were characterized in a manner which he felt to be acceptable to nominalists.

The nominalistic tradition has been carried on by many other Polish logicians, among which Leon Chwistek should be mentioned. Chwistek was willing to endorse only things, matter, sensations, and images. He constructed a semantical theory in which all relations of designation are avoided, and where expressions are the values of variables.

Still more recently, nominalistic tendencies have been displayed, or programs which are thought to be of nominalistic interest have been undertaken by Nelson Goodman, W. V. Quine, R. M. Martin, J. H. Woodger, and in certain respects also by Alfred Tarski, Leon Henkin, J. R. Myhill, and others.

Although it is difficult to discover views common to all contemporary nominalists, the following appear to be widespread: (1) criticism of the notion of a class, in as much as a class differs from the individual whole composed of its elements, (2) a refusal to postulate the existence of an infinity of objects, (3) an aversion to treat predicates as expressions which designate non-individuals, (4) objections to the use of such entities as concepts, meanings, senses, and propositions in the theory of meaning, (5) preference for a syntax where expressions are construed as non-repeatable inscriptions, (6) efforts to reconstruct or reinterpret portions of mathematics in such a fashion that reference to numbers or classes is replaced by reference to concrete objects, individual inscriptions, or wholes which are determined by their least parts, (7) advocacy of parsimony with regard to the number of distinct categories of entities to which a theory makes reference, even at the expense of greater complexity in the construction of definitions and proofs, (8) a tendency to identify individuals, if possible, with phenomenal data or with observable macroscopic things or events. Concrete things are preferred to abstract items, actual entities to possible ones, occurrent qualities to dispositional ones, and observables to theoretical constructs.

Among the types of approaches which have recently been taken to the old problem of universals, and the kinds of distinctions which have been made between universals and individuals, we mention especially the following:

(1) Some authors have tried to clarify the differences between individuals and universals by separately formalizing certain axiomatic systems which were felt to generate the necessary truths peculiar to the notion of an individual or those characteristic of some notion of a universal. For example, if universals are regarded as classes and individuals as non-classes, a formal distinction can be made by constructing appropriate calculi of classes and calculi of individuals, and by contrasting the theorems peculiar to these calculi. This program of clarifying the distinction between individuals and non-individuals is analogous to that of specifying the difference between natural numbers and real numbers by constructing theories whose theorems are regarded as analytic of the corresponding numerical notions. Once such systems are constructed, one can profitably compare a system of individuals with a system of classes with respect to their relative expressive and deductive powers,

their simplicity, intuitive plausibility, suitability as foundations for various theoretical purposes and applications to science.

(2) Frequently, contemporary nominalism takes the form of objections to specific formal devices which are felt to give rise to some suspicious entities. For example, the compression of all objects which have a certain property into a set which is uniquely determined by that property may seem to open a door to Platonic heaven. Others have felt that the move from discourse concerning entities which are equivalent in some respect to discourse concerning equivalence classes of such entities is the one which typically introduces abstract entities. Thus, the operation of abstraction may seem to be typically used when assertions regarding congruent numbers are replaced by ones regarding congruence classes of numbers, or when statements about individuals which are indistinguishable from the same things give way to ones about quality classes of individuals. Again, some nominalists have felt that they could endorse any items which are nameable, describable, or definable by some expression of a language, but they rejected the assumption of non-specifiable entities. Others have objected to certain existential axioms of set theory, notably to the axioms of comprehension, of infinity, and of choice, either because they seemed implausible revisions supplanting contradictory axioms, or because they appeared to assert the existence of totalities which cannot be construed as concrete manifolds. Although objections to specific assertions of mathematics can hardly be regarded as a nominalistic theory, they are often backed up by arguments which lend themselves to generalization and which might give rise to some comprehensive and systematic position regarding universals.

(3) From a different point of view, individuals and universals have been distinguished with respect to certain empirical relations which entities in one of these categories enter or fail to enter. Thus, individuals may bear certain spatial and temporal relations to one another which are felt to distinguish them from non-individuals. Or, individuals may enter certain distinctive epistemological relations: individuals, it is frequently said, are perceived, whereas universals are conceived. Universals, one may hold, result from some activity of abstraction from features irrelevant to a given purpose, and are distinguished from individuals by the manner in which they are arrived at, or used, or serve the purposes of human beings who engage in some inquiry.

(4) A functional distinction between individuals and universals has been sought in the role played by predicates, common nouns, or other types of expressions in ordinary or philosophic discourse.

(5) Some philosophers distinguish universal and individual entities with respect to the function performed by these entities in the semantical interpretation of formal languages. Semantics itself, according to Quine's well-known distinction, can be divided into the theory of meaning and the theory of reference. Correspondingly, the semantical problem of universals can be understood either (a) as the question whether an adequate treatment of oblique or referentially opaque contexts requires that we assign to expressions such entities as concepts, properties, or propositions, or (b) as the question whether an adequate treatment of extensional contexts requires that we assign to various types of expressions entities belonging to more than one category. A realist, with respect to this issue, may affirm and a nominalist deny that an adequate theory of truth, or satisfaction, or meaning requires that one assigns to certain expressions not only individuals, but also classes of individuals, ordered n-tuples of individuals, functions from universes of individuals into such universes, and so forth.

Contemporary theories of realism are frequently called 'Platonism' – especially by their opponents. We shall avoid this term, since these contemporary doctrines have frequently very little in common with the theory of forms advanced by Plato himself. We shall also avoid classifying theories as 'conceptualistic', since it is not clear what conceptualism amounts to. Indeed, even experts on medieval philosophy appear to disagree when they classify philosophers in this category.

In surveying modern views which are claimed to be nominalistic, one is impressed by the great variety among them. It is difficult to discover family resemblances even among the basic approaches taken by nominalists. We shall attempt to give systematic treatment to some of these views under exclusion of others, without thereby intending to imply that alternative approaches are less worthy of attention.

1.3. FORMAL PRELIMINARIES

In the subsequent chapters extensive use is made, both explicitly and implicitly, of symbolic logic, set theory, and the meta-theory of logic.

However, the topics to be covered are presumed to be of interest to readers whose background in these disciplines may be quite varied. For this reason we shall use symbolism only where it is clearly in the service of precision, and we shall provide an English paraphrase of every symbolic formula. In addition, small print will be used for most proofs or outlines of proofs occurring in the main text, and certain technical sections containing longer proofs will be marked with an asterisk and may be omitted without loss of continuity among the philosophical remarks. In this manner, deductive details which may appear too trivial or too complicated may be skipped. Furthermore, we devote this section to some general remarks and to a summary of some technical notions which may elucidate and simplify later developments.

In setting up the formal theories to be given, the customary distinction between object-languages and meta-languages is observed. The expressions of each theory occur in its object-language, but the names of these expressions and variables ranging over the expressions occur in the meta-language in which the syntax and semantics of the theory is given.

The *meta-language* of the formal theories to be stated shall be English enriched by a few symbols and technical expressions. The deductive apparatus which is informally expressed in the meta-language is that of set-theory. Thus, the theory within which nominalistic systems are discussed is not itself a nominalistic one. The precise content of this set-theory will only rarely be of importance. But to fix our thoughts and to facilitate formal reconstructions in an axiomatic system, we agree, until further notice, to employ a Zermelo-Fraenkel set-theory without individuals, but supplied with the axioms of regularity, infinity, and of choice. For convenient reference, a brief outline of some set-theoretic ideas and of some notions pertaining to formal syntax and semantics is provided in the following sections.

1.3.1. *Summary of Elementary Set-Theoretic Notions*

1.3.1.0. $x=y$; read 'x is *identical* with (or, the *same* as) y'. The symbol '$=$' is treated as a logical constant within a theory of identity underlying set theory.[1]

1.3.1.1. *The x such that ϕ*, where ϕ is some formula. Such expressions, called definite *descriptions*, are treated in a logic of descriptions presupposed by set theory. Descriptions of the given form are said to be *proper* if one and only one item satisfies the condition ϕ; otherwise, they are *improper*. Following Frege, all improper descriptions of the metalanguage are identified and, once the empty set is introduced, set equal to that set.[2]

1.3.1.2. $x \in y$; read '*x* is a *member* (or, an *element*) of *y*, or *in y*'. The symbol '$\in$' is the only primitive expression peculiar to set-theory. The negation of '$x \in y$' is '$x \notin y$'.

1.3.1.3. $\{x: \phi\}$, where ϕ is some formula; read '*the set of all objects x such that ϕ*'. We say that $\{x: \phi\}$ *exists* (or, *is proper*) if all and only those items are members of it which satisfy the condition ϕ; otherwise, we say that $\{x: \phi\}$ *does not exist* (or, *is improper*). Unless explicit exception is taken, all sets to be mentioned will be proper.

1.3.1.4. 0; read '*the empty set*' (or, '*zero*'): that set which has no members.

1.3.1.5. $x \subseteq y$; read '*x* is *included* in (or, a *subset* of) *y*': every member of *x* is a member of *y*. $x \subseteq x$; if $x \subseteq y$ and $y \subseteq x$ then $x = y$; if $x \subseteq y$ and $y \subseteq z$ then $x \subseteq z$; $0 \subseteq x$.

1.3.1.6. $x \subset y$; read '*x* is *properly included* in (or, a *proper subset* of) *y*': $x \subseteq y$ and $x \neq y$. It is not the case that $x \subset x$; if $x \subset y$ then not: $y \subset x$; if $x \subset y$ and $y \subset z$ then $x \subset z$.

1.3.1.7. $x \cup y$; read '*x union y* (or, the union of *x* and *y*)': the set of all items which are members of *x* or *y* or both. $x \cup x = x$; $x \cup y = y \cup x$; $x \cup (y \cup z) = (x \cup y) \cup z$; $x \subseteq x \cup y$; $x \cup 0 = x$.

1.3.1.8. $x \cap y$; read 'the *intersection* of *x* and *y*': the set of all items which are members of both *x* and *y*; $x \cap x = x$; $x \cap y = y \cap x$; $x \cap (y \cap z) = (x \cap y) \cap z$; $x \cap y \subseteq x$; $x \cap 0 = 0$; $x \cap (y \cup z) = (x \cap y) \cup (x \cap z)$; $x \cup (y \cap z) = (x \cup y) \cap (x \cup z)$.

1.3.1.9. $x \sim y$; read '*x minus y*' or 'the (set-theoretic) *difference* between *x* and *y*': the set of all items which are members of *x* but not of *y*; $x \sim 0 = x$; $0 \sim x = 0$;

1.3.1.10. The *power set* of *x*: the set of all those sets which are included in *x*. *x* and 0 are members of the power set of *x*.

1.3.1.11. $\{x\}$; read '*singleton x*' or 'the *unit set* of *x*': the set of just those items which are identical with *x*. $x \in \{x\}$; if $y \in \{x\}$ then $y = x$; $x \in y$ just in case $\{x\} \subseteq y$.

1.3.1.12. $\{x, y\}$; read '*doubleton x, y*': the set of just those items which are identical with x or y. Both x and y but nothing else are members of doubleton x, y.

1.3.1.13. $\{x_0, ..., x_{n-1}\}$: the set whose members are exactly the items $x_0, ..., x_{n-1}$.

1.3.1.14. $\bigcup x$; read 'the *union* of x': the set of all those items y which are members of some element of x. $\bigcup 0 = 0$; $\bigcup \{x\} = x$; $\bigcup \{x, y\} = x \cup y$; if $x \in y$ then $x \subseteq \bigcup y$; $\bigcup y$ is the supremum of y relative to inclusion: every member of y is included in $\bigcup y$, and for all z, if every member of y is included in z then $\bigcup y$ is included in z.

1.3.1.15. $\bigcap x$; read 'the *intersection* of x': the set of all those items y which are members of every element of x. $\bigcap 0 = 0$; $\bigcap \{x\} = x$; $\bigcap \{x, y\} = x \cap y$; if $y \neq 0$ then $\bigcap y$ is the infimum of y relative to inclusion: $\bigcap y$ is included in every member of y, and for all z, if z is included in every member of y then z is included in $\bigcap y$.

1.3.1.16. $\{\tau(x_1, ..., x_n): \phi(x_1, ..., x_n)\}$, where $\tau(x_1, ..., x_n)$ is a term of n places and $\phi(x_1, ..., x_n)$ is a formula; read '*the set of all* $\tau(x_1, ..., x_n)$ *such that* $\phi(x_1, ..., x_n)$': the set of all items y such that for some $x_1, ..., x_n$, $y = \tau(x_1, ..., x_n)$ and $\phi(x_1, ..., x_n)$.

1.3.1.17. (x, y); read 'the *(ordered) pair* (or, *couple*) x, y'. Defined as $\{\{x\}, \{x, y\}\}$. $(x, y) = (u, v)$ only if $x = u$ and $y = v$.

1.3.1.18. R is a *relation* just in case for every x in R there exist y, z such that $x = (y, z)$: R is a set of ordered pairs.

1.3.1.19. xRy; read 'x bears R to y': $(x, y) \in R$.

1.3.1.20. The *domain* of R: the set of all items x which bear R to something or other.

1.3.1.21. The *range* of R: the set of all items y to which something or other bears R.

1.3.1.22. The *field* of R: the set of all items x such that either x bears R to something or something bears R to x; the set of all objects which enter (the relation) R.

1.3.1.23. $\check{R}$; read 'the *converse* of R': the set of all items z such that for some x and y, $z = (y, x)$ and $(x, y) \in R$. $x\check{R}y$ just in case yRx; if R is a relation then $\check{\check{R}} = R$.

1.3.1.24. R is *reflexive in* S just in case for every x in S, xRx. R is *reflexive* if R is reflexive in the field of R.

1.3.1.25. R is *irreflexive in* S just in case for every x in S, not: xRx. R is *irreflexive* if R is irreflexive in its field.

1.3.1.26. *R* is *symmetric in S* just in case for all x, y in S: if xRy then yRx; and *R* is *symmetric* if symmetric in its field.

1.3.1.27. *R* is *asymmetric in S* just in case for all x, y in S: if xRy then not: yRx; and *R* is *asymmetric* if asymmetric in its field.

1.3.1.28. *R* is *antisymmetric in S* just in case for all x, y in S: if xRy and yRx then $x=y$; and *R* is *antisymmetric* if antisymmetric in its field.

1.3.1.29. *R* is *transitive in S* just in case for all x, y, z in S: if xRy and yRz then xRz; and *R* is *transitive* if transitive in its field.

1.3.1.30. *R* is a *partial ordering of S* just in case *R* is reflexive, antisymmetric and transitive in S; and *R* is a *partial ordering* if *R* is a partial ordering of its field.

1.3.1.31. *R* is a *strict partial ordering of S* just in case *R* is asymmetric (and hence also irreflexive) and transitive in S; and *R* is a *strict partial ordering* if *R* is a strict partial ordering in its field.

1.3.1.32. *R* is an *equivalence relation in S* just in case *R* is reflexive, symmetric, and transitive in S; and *R* is an *equivalence relation* if *R* is an equivalence relation in its field.

1.3.1.33. The *equivalence class* of x relative to R: the set of all items y such that xRy and R is an equivalence relation. If R is such a relation and x, y are in its field, then xRy if and only if the equivalence class of x is identical with the equivalence class of y (relative to R).

1.3.1.34. x is an *R-minimal element of S* just in case $x \in S$ and there exists no y in S such that yRx.

1.3.1.35. R_S; read '*R restricted* to *S*', or 'the *restriction* of *R* to *S*': the set of all pairs (x, y) in R where both x and y are in S.

1.3.1.36. Id_S; read 'the *identity relation restricted to S*': the set of all pairs (x, x), where $x \in S$.

1.3.1.37. $\subseteq_S$; read '*inclusion restricted to S*': the set of all pairs (x, y) which are such that $x \subseteq y$ and $x, y \in S$.

1.3.1.38. f is a *function* just in case f is a relation and for all x, y, z, if xfy and xfz then $y=z$. 0 is a function; if f is a function and $g \subseteq f$ then g is a function; if f and g are functions and the intersection of their domains is empty then $f \cup g$ is a function. We write $f(x)$, read 'the *value of f* evaluated at x' for that object y which is such that xfy (if there is a unique such y; $f(x)=0$, otherwise). Two functions f and g are identical just in case their domains are the same and for every x in the domain of $f, f(x)=g(x)$.

1.3.1.39. f is 1-1; read 'f is a *one-to-one* function' just in case both f and the converse of f are functions. 0 is 1-1; if f is 1-1 then $\check{f}$ is 1-1; if f is a function and for all x and y in the domain of f, if $f(x)=f(y)$ then $x=y$, then f is 1-1. If f is 1-1 and x is in its domain, then $\check{f}(f(x))=x$. Id_S is 1-1.

1.3.1.40. f/g; read 'the *relative product* of f and g': the set of all ordered pairs (x, y) such that for some z, xfz and zgy. f/g is a relation and its domain is included in the domain of f; if both f and g are functions (or 1-1) then f/g is a function (or 1-1) and for every x in its domain, $(f/g)(x)=g(f(x))$.

1.3.1.41. x is *equinumerous* with y just in case there exists a 1-1 function f whose domain is x and whose range is y. x is equinumerous with x; if x is equinumerous with y then y is equinumerous with x; if x is equinumerous with y and y with z, then x with z.

1.3.1.42. x has *at most as many elements as* y just in case x is equinumerous with some subset of y.

1.3.1.43. x has *fewer elements than* y just in case x has at most as many elements as y and is not equinumerous with y.

1.3.1.44. x is *denumerable* just in case x is equinumerous with the set of *natural numbers* (which will not be characterized here); that is the non-negative integers.

1.3.1.45. x is *finite* just in case for some natural number n, x is equinumerous with the set of all natural numbers less than n. A set which is not finite is said to be *infinite*. There exists an infinite set (which is the content of the axiom of infinity); the set of all natural numbers is infinite; and every infinite set is equinumerous with a proper subset of itself.

1.3.1.46. x has the *power of the continuum* if x is equinumerous with the power set of some denumerable set. Denumerable sets have fewer elements than ones which have the power of the continuum. The set of all real numbers has that power.

1.3.1.47. f *is an n-term sequence* just in case f is a function whose domain is the set of all numbers less than n. f is a *finite sequence* if for some natural number n, f is an n-term sequence. A 0-term sequence is 0.

1.3.1.48. $\langle x_0, \ldots, x_{n-1} \rangle$; read '*the n-term sequence* whose respective terms are $x_0, \ldots, x_{n-1}$': the n-term sequence x (for every number i less than n, we write 'x_i' for '$x(i)$'). The 1-term sequence $\langle x \rangle$ is that function whose unique element is $(0, x)$; the 2-term sequence $\langle x, y \rangle$ is that function whose domain is $\{0, 1\}$, which assigns x to 0 and y to 1; if $\langle x_0, ..., x_{n-1} \rangle = \langle y_0, ..., y_{m-1} \rangle$ then $m = n$ and for all i less than n, $x_i = y_i$. We shall often disregard the differences between 2-term sequences and ordered pairs.

1.3.1.49. *R* is an *n-term* (or, *n-adic*) *relation in U* just in case *R* is a set all of whose members are *n*-term sequences whose range is included in *U*; and *R* is an *n-term* (or, *n-adic*) *relation* if all members of *R* are *n*-term sequences. Where the difference does not matter, we shall speak interchangeably of relations (sets of pairs) and of 2-term (or diadic) relations (sets of sequences). 0 and $\{0\}$ are the only 0-term relations. If *R* is a 1-term relation in *U* then all members of *R* have the form $\langle x \rangle$ for some x in *U*.

1.3.1.50. *f* is an *n-term operation in U* just in case *f* is a function whose domain is an *n*-term relation in *U* and whose range is included in *U*; *f* is an *n-term operation on U* if *f* is an *n*-term operation in *U* whose domain is the set of all *n*-term sequences whose terms are in *U*; *f* is an *n-term operation* if *f* is a function whose domain is an *n*-term relation. If *f* is a 0-term operation on *U*, then $f(0)$ is in *U*. If all of the items $x_0, ..., x_{n-1}$ are in *U* and *f* is an *n*-term operation on *U*, then $f(\langle x_0, ..., x_{n-1} \rangle)$ is again a member of *U*.

1.3.1.51. *S* is *closed under the relation R* just in case *R* is a relation and whenever x is in *S* and xRy then y is again in *S*; and *S* is *closed under the n-term operation f* if *f* is an *n*-term operation on *S*.

1.3.1.52. x is an *ancestor, with respect to R, of y* just in case x is a member of every set *S* such that y is in *S* and *S* is closed under the relation *R*; and x is a *proper ancestor, with respect to R, of y* if x is in every set *S* which is such that everything which bears *R* to y is in *S* and *S* is closed under the relation *R*. The *ancestral of R* (respectively, the *proper ancestral of R*) is the set of all ordered pairs (x, y) such that x is an ancestor (proper ancestor), with respect to *R*, of y. x is a *proper ancestor, with respect to membership, of y* if x is in every set *S* which is such that every member of y is in S ($y \subseteq S$) and whenever u is in S and $v \in u$ then $v \in S$ ($\bigcup S \subseteq S$). Ancestrals are transitive.

1.3.1.53. *R* is *isomorphic with S* if there exists a one-to-one function *f* whose domain is the field of *R*, whose range is the field of *S* and such that, for all x and y in its domain, xRy if and only if $f(x) Sf(y)$.

For elaborations and further set-theoretic ideas we refer to Suppes (1960).

1.3.2. *Formal Systems, their Syntax and Semantics*

The symbols occurring in the object-language of every formal system to be considered are drawn from the following categories:

(a) a denumerable set of *individual variables*: lower case Latin letters with or without numerical subscripts, of which 'x', 'y', 'z', 'u', 'v', and 'w' occur most frequently;

(b) the two parentheses '(' and ')';

(c) the five *sentential connectives* '$\sim$', '$\rightarrow$', '&', '$\vee$', and '$\leftrightarrow$', which are the respective symbols of negation (read 'it is not the case that'), conditional (read 'if... then ---'), conjunction (read 'and'), disjunction (read 'or'), and biconditional (read '... if and only if ---');

(d) the universal quantifier '$\forall$' (read 'for all ...') and the existential quantifier '$\exists$' (read 'for some ...');

(e) for every natural number n, there is an infinite set of *n-place operation symbols* which we identify officially with the Capital Latin letters 'A' through 'E' with or without numerical subscripts and bearing the superscript 'n'. Superscripts may be omitted where the number of places is clear; and further operation symbols may be introduced by definition (e.g., the two-place operation symbol '$+$') without explicitly identifying the new symbols with Capital letters. Zero-place operation symbols (or *individual constants*) are symbolic names whose English analogs comprise denoting expressions such as 'Rudolf Carnap' and 'the Eiffel Tower', non-denoting names such as 'Zeus' and 'Rumpelstiltkin', and names whose referential status is disputed such as 'Two' and 'Whiteness'. Operation symbols of n places have English counterparts in expressions which, when suitably complemented by n names, become themselves names; such as 'the father of...', 'the sum of... and ---', 'the result of replacing... by --- in = = =', and so forth.

(f) for every positive integer n, there is an infinite set of *n-place predicate symbols*: the Capital Latin letters 'F' through 'Z' with or without numerical subscripts but supplied with the superscript 'n' which, in clear contexts, may be omitted. Further predicate symbols may be defined. One-place predicate symbols (or *predicates*) are formal analogs of English predicate or verb phrases, such as 'is blond' and 'reads a book'. Predicate symbols of two and more places (or *relation expressions*) may abbreviate

English phrases like '... is taller than ---', '... resembles --- with respect to = = =', and so on.

No symbol is common to two or more of these syntactical categories; and no symbol is regarded as a concatenate of two or more symbols, even if it should display typographical complexity. Since set theory is embodied in the meta-language, we can be assured of an infinite stock of entities which may serve as symbols. In connection with each system to be given, we shall specify the particular subsets of these categories of symbols which are to occur in the language of the system.

Given the symbols of a system, two sorts of well-formed expressions will be characterized in the customary manner: *formulas* (which are sentences or expressions which result from sentences by replacing names by variables) and *terms* (that is, names or expressions which result from names by replacing names occurring in them by variables). The grammatical notions of *freedom* and *bondage* of variables and of occurrences of variables and that of *alphabetic variance* will be standard and their definitions presupposed.[3]

The following syntactical terminology and notation will be used throughout:

1.3.2.1. $\ulcorner \xi_0 \ldots \xi_{n-1} \urcorner$; read 'the *concatenate* of the respective expressions $\xi_0, \ldots, \xi_{n-1}$'. We use the *corners* or quasi-quotation marks in the manner of Quine[4] to indicate the result of concatenating (or juxtaposing) those expressions which are the referents of the constants or the values of the variables displayed within the quotes, each symbol of the object-language being its own name. *Display* on separate lines may be used in the same sense as corners. In informal contexts, ordinary quotation marks may be used both to form the standard name of an expression and in the stead of corners; and in such contexts, syntactical generality may be indicated by the words 'an expression of the form...' instead of by variables ranging over expressions.

1.3.2.2. *Greek letters* are variables of the meta-language whose values are expressions of the object-language. Unless otherwise indicated, the variables 'α', 'β' and 'γ' range over variables; 'ϕ', 'ψ', and 'χ' over formulas; 'τ', 'ζ', and 'η' over terms; 'o' over operation symbols, and 'π' over predicates.

1.3.2.3. $\phi_{\tau_0 \ldots \tau_{n-1}}^{\alpha_0 \ldots \alpha_{n-1}}$; read 'the *proper substitution* of $\tau_0, \ldots, \tau_{n-1}$ respectively for $\alpha_0, \ldots, \alpha_{n-1}$ in ϕ'. It is the result of replacing every free occur-

INTRODUCTION   19

rence of the variable α_i by a free occurrence of the term τ_i (for all $i<n$) in some alphabetic variant ϕ' of ϕ (ϕ' being uniquely determined in some manner which is not important here).

The *logical axioms* of all theories will be ones appropriate to a first-order predicate calculus which allows for denotationless terms and for the empty universe of discourse. Since the logical axioms deviate slightly from the customary ones, they will be explicitly stated as part of the first system to be formulated.

1.3.2.4. For *inference rules* we shall employ Modus Ponens and Universal Generalization on the consequent of conditionals. Formally, these rules are expressed in the following closure principles: A set K is said to be *closed under modus ponens* just in case ψ is a member of K whenever both ϕ and $\ulcorner(\phi \to \psi)\urcorner$ are members of K. A set K is said to be *closed under universal generalization* just in case $\ulcorner(\phi \to \forall\alpha\psi)\urcorner$ is a member of K whenever $\ulcorner(\phi \to \psi)\urcorner$ is a member of K and α is a variable which is not free in ϕ.

1.3.2.5. By the *vocabulary of* a *term* or *formula* we mean the set of predicates and operation symbols occurring in that term or formula.

1.3.2.6. A *formal system* is to be an ordered pair $\langle V, L \rangle$, where V (the *vocabulary of* the *system*) is a set of predicates and operation symbols, and L (the set of *proper axioms* of the system) is a set of formulas each member of which has its vocabulary included in V.

Under the convention that $\langle V, L \rangle$ is to be a formal system, we say:

1.3.2.7. ξ is a *term* or *formula of* $\langle V, L \rangle$ if ξ is a term or formula whose vocabulary is included in V.

1.3.2.8. ϕ is a *theorem of* $\langle V, L \rangle$ if ϕ is a formula of $\langle V, L \rangle$ and ϕ is in every set K which is such that all logical axioms are in K, L is included in K, and K is closed under modus ponens and under universal generalization.

1.3.2.9. ϕ *is derivable from* K *in* $\langle V, L \rangle$ if ϕ is a formula of $\langle V, L \rangle$, K is a set of formulas of $\langle V, L \rangle$, and ϕ is in every set S which is such that K is included in S, all theorems of $\langle V, L \rangle$ are members of S, and S is closed under modus ponens.

1.3.2.10. K is *consistent in* $\langle V, L \rangle$ if K is a set of formulas of $\langle V, L \rangle$ and there is no formula ϕ of $\langle V, L \rangle$ such that both ϕ and $\ulcorner\sim\phi\urcorner$ are derivable from K in $\langle V, L \rangle$.

1.3.2.11. K is *maximally consistent in* $\langle V, L \rangle$ if K is consistent in

$\langle V, L \rangle$ and no set of formulas of $\langle V, L \rangle$ properly including K is consistent in $\langle V, L \rangle$.

With respect to every formal system, the customary distinction between *primitive* and *defined* symbols is observed. The division of the vocabulary into primitive and defined symbols satisfies, in outline, the following conditions: each defined symbol is to occur appropriately in exactly one axiom of the system (called a *definitional axiom*) and the axiom is an explicit definition of the symbol in terms of the primitive symbols of the theory.[5] Since definite descriptions will not be introduced into the object-languages, all definitions, including definitions of terms, will have the form of biconditionals. Since the purpose of defining symbols at all is to facilitate the reading and intuitive comprehension of formulas, in practice some defined symbols will occur in the axioms which are taken as definitions of other defined symbols. In all cases it will be obvious that no circularity is involved. Due to the properties of definitions, inductive arguments required in proving results about formal systems can be simplified by taking account only of terms and formulas containing primitive symbols exclusively and of axioms other than definitional ones. Also, on the basis of known properties of the underlying logic, we need bring into consideration in such proofs only the symbols of negation, conditional and universal quantification.

One method of characterizing the *semantics* of a formal system is basically due to Tarski. As a first step, one specifies the admissible *models* of the system, each model providing a possible interpretation of the primitive non-logical symbols relative to some universe of discourse. We define one variant of this notion which, in distinction from other conceptions of a model to be discussed, we call a 'classical model':

1.3.2.12. If $\langle V, L \rangle$ is a formal system, then a *classical model* for $\langle V, L \rangle$ is an ordered quadruple $\langle \mathbf{U}, \mathbf{O}, \mathbf{P}, \mathbf{A} \rangle$ meeting the following conditions: (1) $\mathbf{U}$ (the *universe of discourse* of the model) is any non-empty set; (2) $\mathbf{O}$ (the *operation assignment* of the model) is a function whose domain is the set of all operation symbols in V, and for every such symbol o of n places, $\mathbf{O}(o)$ is an n-term operation on $\mathbf{U}$ (1.3.1.50); (3) $\mathbf{P}$ (the *extension assignment* of the model) is a function whose domain is the set of all predicate symbols in V, and for every such symbol π of n places, $\mathbf{P}(\pi)$ is an n-term relation in $\mathbf{U}$ (1.3.1.49); (4) $\mathbf{A}$ (the *assignment to variables* of the model) is a function whose domain is the set of all variables and whose values are always in $\mathbf{U}$.

If we pretend that English expressions are in the vocabulary of some formal system and that the universe **U** of such a model is the class of persons, we can informally explain how the intended interpretation takes place. The operation assignment **O** is a rule which may assign to the name (or, zero-place operation expression) 'Tom' the person Tom (more accurately, $O(\text{'Tom'})(O) = \text{Tom}$); it may assign to the one-place operation expression 'the father of' an operation which, when applied to any person in **U** (more accurately, to the 1-term sequence of that person) gives the father of that person; and it may assign to the expression 'the senior of ... and ---' an operation which associates with every sequence of two persons the elder among the two persons. The extension assignment **P** is a rule which may assign to the predicate 'is pale' the class of all pale persons in **U** (more accurately, the set of all 1-term sequences of such persons); it may associate the 2-term relation expression 'is taller than' with the class of all 2-term sequences $\langle x, y \rangle$ where x and y are persons in **U** and x is taller than y (that is, with the diadic relation 'taller than' restricted to the universe **U**); and so forth. The assignment **A** specifies for each variable which person in **U** shall be regarded as the value of that variable.

There will be occasion to speak of models in a wider sense. For example, a class of models for a formal system may be specified by the condition that all universes of discourse of such models shall be subsets of a given reference class. Such reference classes may be ordered with respect to one or several relations, and various conditions may be imposed on them. Models whose universe of discourse is empty may also be admitted.

The operation and extension assignments of models in a broad sense may be eccentric in various respects: their domains may comprise symbols or also sequences some constituents of which are symbols and others are elements of the universe of discourse. The assignments in question need not interpret the entire vocabulary of a system, leaving some symbols 'denotationless'. The values of these assignments may be members of the universe rather than operations or relations in the universe.

Models, in the broad sense, may also have constituents other than the ones mentioned in connection with classical models. The precise nature of such additional constituents will be made explicit when they are introduced.

In the course of our discussion, various notions of models will be explored, not so much with a view to their general algebraic properties, but rather with the aim of assessing their suitability as tools for interpreting formal systems. We shall not attempt to provide a rigorous definition of 'model' in a sense wide enough to cover all of the numerous conceptions to be explored. For such a definition would be quite artificial and would tend to obscure, rather than aid, our understanding of various alternative proposals.

Once the admissible models of a system have been specified, one makes explicit, by clauses appropriate to well-formed expressions of increasing complexity (that is, by a recursive definition), what each term of the system is to *denote* (or, have for its *value*) in a model, and under what conditions a formula is *satisfied* (or, a sentence is *true*) in the model. Reverting to the informal example given earlier, the name 'Tom' may be said to denote in the classical model $\langle \mathbf{U}, \mathbf{O}, \mathbf{P}, \mathbf{A} \rangle$ the person assigned by $\mathbf{O}$('Tom') to the empty sequence; and 'the father of Tom' may denote the result of applying the operation assigned by $\mathbf{O}$ to 'the father of' to the sequence whose unique term is the person denoted by 'Tom'. Thus, intuitively, 'the father of Tom' denotes the father of Tom. And the sentence 'Tom is pale' may be true in that model if the person denoted by 'Tom' (or rather, the sequence whose unique term he is) is a member of the extension assigned by $\mathbf{P}$ to the predicate 'is pale'. And the sentence 'Tom is taller than the father of Tom' may be true under the given interpretation if the person denoted by 'Tom' bears to the person denoted by 'the father of Tom' that diadic relation which is assigned, by $\mathbf{P}$, to 'is taller than'.

A formula of a system is said to be *valid* or *logically true* if it is true in every admissible model of the system. Given a formal system, a class K of formulas of the system, and a formula ϕ of the system, we say that K *semantically yields* ϕ (or that the argument whose premises are the members of K and whose conclusion is ϕ is a *valid* one) just in case the following condition is met by every model of the system: if every formula in K is satisfied in the model, then ϕ is also satisfied in the model. These ideas will be expressed more precisely in the framework of particular systems.

It is customary to distinguish *schematic* and *non-schematic* predicate and operation symbols. Suitable substitution, in theorems, on schematic symbols yields new theorems. This is not true of non-schematic symbols. Non-schematic symbols may occasionally be called *logical constants* in order to indicate an intention to keep the interpretation of such symbols invariant within a given system (that is, to interpret them by means of satisfaction conditions appropriate to the formulas in which they occur, rather than by means of operation or extension assignments in models).

We say that a *model-theoretic interpretation* has been provided for a formal system when a set of models has been specified and a definition of satisfaction given. Once a system and a model-theoretic interpretation are available we speak of the *limited soundness* of the system if every theorem of the system is valid; and we say that the system is *sound* if the following condition is satisfied by every set K of formulas of the system and by every formula ϕ of the system: if ϕ is derivable from K in the system, then K semantically yields ϕ in the system. Under similar assumptions, we credit a system with *limited completeness* if every valid formula of the system is a theorem of the system; and we say that the system is *complete* if the following condition is met by every class K of formulas of the system and by every formula ϕ of the system: if K semantically yields ϕ in the system, then ϕ is derivable from K in that system. An interpreted formal system which is both sound and complete satisfies a condition of *semantical adequacy*. With regard to the formal systems whose semantical adequacy will be demonstrated, we shall usually undertake to prove their limited soundness only, since their soundness is easily obtained from their limited soundness. On the other hand, we shall usually prove their completeness, of which their limited completeness is an immediate corollary.

REFERENCES

[1] See, e.g., Kalish and Montague (1964), Ch. VI.
[2] *Ibid.*, Ch. VII.
[3] See, e.g., Kalish and Montague (1964), Ch. VIII, sections 1 and 2.
[4] Quine (1940), Ch. 1, section 6.
[5] See Mates (1965), Ch. 11, section 5.

CHAPTER 2

INDIVIDUALS

2.1. OUTLINE OF GOODMAN'S CONCEPTION

Nominalism, according to Nelson Goodman, consists in a refusal to countenance any entities other than individuals. Hence, in order to clarify this position of nominalism, the notion of an individual must be explicated. Goodman's insights concerning that notion have been more influential and probably clearer than any other account of individuals which is relevant to the distinction between nominalism and realism. His explication of the notion has taken two forms: (1) the construction of a certain formal system, called a 'calculus of individuals', which will be examined in a subsequent section, and (2) informal and semi-technical remarks regarding the difference between individuals and classes and the consequences of endorsing nothing but individuals. In order to cultivate our intuitions, we begin with a brief outline of the latter, inserting more detailed discussions at appropriate stages of our later developments.

Goodman is aware that a nominalist's decision to recognize nothing but individuals does not of itself determine what sorts of entities may be taken as individuals. He allows that individuals may be abstract or concrete, particular or universal, physical or phenomenal.[1] Indeed, any one entity of whatever kind may be construed as an individual.[2]

Goodman concedes, furthermore, that any individual can be construed as a class. This can be done, e.g., by identifying physical objects with the classes of their atomic or sub-atomic parts or with certain classes of physical events. One may also construe individuals as classes by translating discourse concerning individuals into discourse concerning their unit sets.

An important concept in Goodman's program is that of a 'sum' or whole composed of several individuals. Any whole of individual parts is always again an individual, whether or not the parts of such a whole lie close together in space or time. Thus, though red or round individuals may occur at widely separate places and moments, Goodman will nevertheless admit as an individual the whole which is composed exactly of all red individuals or of all round individuals, or of either. There is an individual whole comprising exactly the three individuals Julius Caesar, the Empire State Building, and the North Star.

On Goodman's view, the general theory of individuals differs from the general theory of classes.[3] But individuals are not distinct from classes by virtue of being made up of a different and special kind of 'stuff'. Nor is the criterion for the distinction an epistemological one. It is not claimed, e.g., that individuals can be perceived, whereas classes cannot; or that classes are mentally constructed, whereas individuals are 'given'. And it is not the fact that classes can have elements which, in general, distinguishes them from individuals. Furthermore, it seems that the principle of 'sum formation' is quite analogous to the set-theoretic principle governing unions of singletons. For example, the set of all red objects (the union of all singletons of red objects) has for its counterpart a certain individual, namely the sum of all red objects. What then is the basis for distinguishing classes and individuals?

An example, modified from Quine's[4] may clarify somewhat the intended distinction between individuals and classes. With regard to a given heap of stones, the following two classes can be distinguished: (1) the class of stones in the heap, and (2) the class of molecules of stones in the heap. Since there are more molecules than stones in the heap, the two classes are certainly distinct. On the other hand, a nominalist's intuition regarding individuals is such that he would not wish to distinguish the following corresponding individuals: (1) the concrete whole composed of all stones in the heap, and (2) the concrete whole composed of all molecules of stones in the heap (in the given spatial arrangement of molecules). Concrete individuals wholes, it is said, which have identical 'ultimate constituents' in identical spatial order, must themselves be identical. Suppose that molecules are the smallest physical objects under consideration, so that for the purposes of this example molecules are the 'ultimate constituents' of physical objects. Then physical wholes which have exactly the same molecules for parts, such as the two heaps in question, should be regarded as identical. Thus, the need for a principle of individuation peculiar to individuals begins to emerge.

Another example may illustrate more clearly the intuitive difference between sameness of classes and sameness of individuals: Given an object without members, say A, the standard principles of set formation imply the existence of a unit class $\{A\}$ of A, of a unit class $\{\{A\}\}$ of the unit class of A, and so on *ad infinitum*. By the principle of extensionality, all of these successive sets are distinct, even though they are built up from what is regarded as just one 'ultimate constituent', the object A. Goodman's conception of an individual precludes that on the basis of just one individual a variety of new individuals can be generated.

Furthermore, given two objects without members[5], say A and B, set theory permits the construction and implies the difference of the two items $\{\{A\}, \{A, B\}\}$, and $\{\{B\}, \{B, A\}\}$ which may be identified, respectively, with the ordered pair (A, B) and the ordered pair (B, A) (see 1.3.1.17). Both of these entities have the same 'content' in the sense that they are both built up from the same objects A and B. What distinguishes the given ordered pairs is just the arrangement among the objects A and B, or the 'structure' of the corresponding sets: the manner in which A and B are grouped together. If we adopt for a moment the slippery terminology of 'content' and 'structure', we can say that individuals which have identical content should be regarded as the same; or, conversely, that entities which differ only with respect to their structure are non-individuals.

If we restrict attention to 'constituents' of collections relative to the proper ancestral of membership (see 1.3.1.52), then the principle of extensionality differentiates classes which have different 'immediate constituents' relative to membership-chains; that is to say, classes which have different members. Individuals which have the same content are to count as identical; and 'having the same content' is here taken to mean 'having the same ultimate constituents'. Thus, if collections are to count as individuals, then the principle of extensionality should be replaced by a different principle of individuation according to which just those collection are identical which have the same 'ultimate constituents'.

Since membership-chains are not the only relations with respect to which individuals can have constituents, it is desirable to generalize the notions of a 'constituent' and of an 'ultimate constituent', as well as to render them more precise. Goodman partially meets these demands by introducing the concept of a *generating relation*. This notion (in 1956) is not defined, but exemplified by the relations of 'being a proper part' (to which axiomatic treatment is given in the calculus of individuals) and by the ancestral of membership. For the sake of clarity and generality, let a generating relation R be a strict partial ordering of a set S (that is, R is asymmetric and transitive in S; see 1.3.1.31). The R-minimal elements of S (that is, those elements of S to which no elements of S bear the relation R; see 1.3.1.34) are called 'atoms' (relative to R) of S. The desired principle of individuation, restricted to non-atomic individuals in S, will then read:

(2.1.1) x is the same as y just in case all and only those atoms of S which bear R to x bear R to y.

In Sections 2.2 and 2.3 such a principle of individuation will be characterized more precisely. But in the framework of the present informal discussion at least the following points should be noted:

(1) Not every strict partial ordering and not every relation which 'generates' in the sense of ordinary discourse can qualify as a generating relation or as a part-whole relation in the sense intended by Goodman. For example, if the individuals under consideration are human beings, then the relation of being a biological offspring, though a strict partial ordering in the field of human beings, is not a suitable generating relation. For human beings need not be identical if they have the same ancestors.[6] It may be indicative of the fact that this relation is not one intended by Goodman, that in ordinary discourse one does not speak of the ancestors as 'parts' of a human being, or of the ultimate ancestors (if there be such) as the 'content' of a person, although they may be said to 'generate' the person.

Furthermore, not everything which might be termed a 'part-whole relation' in customary discourse can be taken as a 'generating' relation in Goodman's sense. For example, even though a hand is properly called a 'part' of a human being, the relation 'being a hand of' is not one relative to which human beings are 'wholes'. The ultimate constituents relative to 'being a hand of' are hands (since hands do not have hands). Two men might lack hands altogether, so that every ultimate constituent, with respect to the given relation, which is part of the one man is part of the other. Yet the individual men would not be identified.

These examples suffice to show that there is a need for further clarification of what may be meant by a generating or part-whole relation.

Goodman, made aware of such difficulties, proposed to restrict the notion of a generating relation essentially just to the ancestral of membership and to the relation of being a proper part which is formally treated in the calculus of individuals.[7] Since the intended interpretation of the calculus of individuals has not been made fully explicit, it may be difficult to assess whether this restriction amounts to a loss of generality. Part of our motivation for investigating the semantics of such calculi is that of permitting more definite answers to such questions.

(2) Although it is frequently said that a nominalist may acknowledge the existence of heaps, though not that of classes, the relevant distinction between heaps and classes is not as clear as it might at first seem. For the difference pertinent to a Goodman-type nominalist is not this: that heaps whose parts have been widely scattered are no longer heaps, whereas classes whose elements have been dispersed remain classes. For, as we have seen, a nominalist may well endorse individual wholes whose parts are scattered.

In ordinary discourse, the expressions 'a heap of stones' and 'a heap of molecules of stones' have distinct meaning. If one were to break up stones into their molecular parts and pile up the resulting dust, one would be left with a heap of molecules of stones. Such a heap has no stone for a part. Imagine a heap A of stones ground up into a heap B of molecules of stones without loss or addition of a single molecule. Normally, one would want to say that the heaps A and B have the same molecules for parts. Molecules might be regarded as ultimate physical constituents for the purposes of this discussion. But then, by the principle of individuation proposed by Goodman, it would seem that the heaps A and B should be regarded as the same individual, though they are not the same heaps. Hence heaps, it would seem, are not individuals; or, if heaps are individuals, Goodman's principle seems far too strong to apply to individuals.

The important difficulty which has been raised here is this: that the *arrangement* among the molecules by virtue of which, in ordinary language, we distinguish a heap of molecules of stones from a heap of stones (molecules arranged in patterns constituting stones) can not serve to differentiate individuals in Goodman's sense. However, at least in informal philosophical discussions, this difficulty is not insurmountable. With respect to the given example, it might be replied that molecules of stones are not suitably chosen 'ultimate parts'. Molecules are such that the same molecules can be said to enter into different arrangements at different times. The difficulty disappears if instead of molecules one regards temporal slices of molecules, or 'molecule-stages', as ultimate parts. Under this new choice of constituents, the earlier heap A and the later heap B would have different 'ultimate parts' and would therefore be different. Alternatively, one might continue to regard the molecules of the two heaps as their 'ultimate constituents', but instead construe the 'generating relation' as being different in the two cases. The heap A of stones, it might be said, is a whole of molecules relative to the relation of being a part at time T_1, whereas the result B of grinding up these stones is a whole of the same molecules, but a whole relative to the relation of being a part at time T_2. In general, whenever the *order* among ingredients would be regarded as differentiating individual wholes, one must either construe the notion of an 'atom' so that different order implies different 'atoms', or else the generating or part-whole relation must be so conceived that different order implies a different generating relation.

Goodman's theory of qualia [8] illustrates how a suitable choice of atoms serves to reconcile ones intuition regarding concrete wholes with the demands of the principle of individuation (2.1.1). Leaving aside features of his theory which are inessential to our present concerns, his treatment of concrete things can be described as follows: Certain qualitative atoms (qualia) mark phenomenal spatial and temporal position (they are position-qualities). Concrete wholes differ with respect to spatial or temporal position just in case they have different position-qualia as parts. In this manner, the demand that it should be possible for concrete individuals to differ just with respect to their spatial or temporal positions is reconciled with the principle of individuation which requires that different individuals have different ultimate constituents.

These examples indicate that a nominalist endorsing such a principle of individuation is inclined to subscribe to a philosophical outlook according to which the 'ultimate' individuals, those with respect to which all individuals are wholes, result upon slicing up reality in many and unusual ways. Thus, rivers may be conceived not only as wholes composed of droplets, but further as composites of temporal stages of droplets, or even of discernible qualities of such stages. Indeed, the 'ultimate' individuals will be 'parts', simultaneously, with regard to at least as many principles of division as there are respects in which individuals are differentiated. For if two individuals differ in any respect R, the given principle of individuation demands that there exists a constituent, selected in accord with R, which is part of one but not the other of the two individuals. As we have briefly observed, there is an alternative to this extreme fragmentation: Instead of introducing new entities with every new respect of differentiation, one may leave given entities alone and multiply instead the part-whole relations in such a fashion that an individual is conceived as a 'whole' of the same given entities with respect to as many part-whole relations as there are respects of differentiation. However, as far as I can tell, this second option has not been exploited by any major philosopher.

(3) One example of an admissible part-whole relation appears to be that obtaining at a given moment between a physical part and a physical whole. Can we be assured that there are indeed 'ultimate constituents' in the whole field of physical objects relative to this relation? Suppose that physical objects turn out to be infinitely divisible; should we then be prepared to admit that physical objects are not individuals? Goodman does not preclude, on principle, that there be an infinite number of least physical constituents. Suppose that the physical part-whole relation is conceived in analogy to the relation of inclusion between closed intervals on the real line and the line itself. Then, since there are least closed intervals, the analogy is that physical bodies split into an infinite number of ultimate parts. In this case, infinite divisibility of physical objects offers no difficulties. On the other hand, suppose that the physical part-whole relation is conceived so as to be similar to the relation of inclusion between open intervals on a line and the line itself. Then, since there are no least open intervals, one conceives of physical bodies which have no ultimate parts. A nominalist of the persuasion considered so far could not regard physical objects of this latter kind as individuals. But it seems unlikely that a nominalism of this sort should ever be incompatible with the observational content of a physical theory. In addition, we shall later formulate a principle of individuation in which no reference is made to ultimate constituents and which is still true to the basic insight that individuals differ only if they have different content. In as much as such a principle can be precisely stated, there is at least one clear notion of an individual which does not entail that there exist parts which have no further parts.

(4) Is Goodman's conception of an individual in close agreement with

the presystematic use of the word 'individual' in philosophical discourse? It seems that in such discourse the word 'individual' is used rather loosely. Often no sharp distinction is made between individuals, particulars, concrete things, material objects, corporeal entities, and physical objects. But among the numerous senses given by philosophers to the word 'individual' there is indeed one according to which individuals are determined by their content rather than by their structure. In attempting to substantiate this claim by mentioning historic precedent, we are admittedly on rather shaky grounds. However, history of philosophy is often undertaken with the primary aim of stimulating thought, rather than with that of achieving strict historical accuracy. In this spirit we draw attention to a similarity between Goodman's notion of an individual and that due to Thomas Aquinas.

Aquinas, it may be recalled, maintained that "the principle of individuation is matter"; that individuals (though uniting both matter and form) are identical just in case they comprise the same matter. The connotations of 'form' and 'structure', and those of 'matter' and 'content' seem similar. By 'form' was meant that which determines membership in classes or species; by 'matter' the potentiality of receiving form, or that which can enter various structures. Thus, individuals which are identical with respect to that ingredient which is capable of assuming various patterns or entering various structural relations are to count as the same. Although Goodman's conception is reminiscent of this earlier notion, he has contributed much to its clarification.

Further remarks concerning the intuitive content of Goodman's proposal, notably discussions regarding 'sums' of individuals, individuals which lack atomic parts, and ones which differ with respect to an arrangement of parts shall be postponed until more precise terminology is at our disposal.

2.2. PART-WHOLE RELATIONS

In the preceding section it has been observed that explication of the notion of an individual calls for the formulation of a principle of individuation approximated by (2.1.1); and this task, in turn, requires analysis of certain auxiliary concepts such as that of a generating relation, of an atom, and of some set comprising elements generated by this relation from such atoms. We now turn to an effort of making these notions precise.

One of the concepts central to Goodman's theory of individuals is that of a 'sum' or 'whole' of parts. One speaks of parts and wholes with respect to many kinds of relations. For example, a concrete individual may be regarded as the whole of its qualitative parts, of its physical molecules, or of the events in its history. A set $\{A, B\}$ is a whole composed of the parts $\{A\}$ and $\{B\}$ relative to inclusion, but a whole of the parts A and B with respect to the ancestral of membership. The notion of a 'sum' should be defined quite generally relative to any relation which has *prima facie* claim to being a part-whole relation. A 'sum', in this general sense, may be identified with the supremum of a set relative to an arbitrary partial ordering (see 1.3.1.30). We are acquainted with the special case of a supremum or least upper bound of a set S of numbers relative to the partial ordering 'less than or equal to': it is the least number x which is such that all numbers in S are less than or equal to x. Relative to inclusion, the supremum of a set S is just the union of S (see 1.3.1.14). Generalizing, we define:

(2.2.1)   DEFINITION: $sup_R S$ [the *supremum*, relative to R, of S] = the unique object y which is such that (1) every member of S bears R to y, (2) for all z, if every member of S bears R to z, then y itself bears R to z, and (3) R is a partial ordering. We say that $sup_R S$ *exists* just in case there is one and only one object y such that the conditions (1), (2), and (3) are satisfied.

In appraising this definition, it should be recalled that by a convention mentioned in (1.3.1.1), all improper descriptions denote the empty set 0. Thus, if R is not a partial ordering, or if there is no y satisfying the conditions (1) and (2), then $sup_R S = 0$.

As immediate consequences of the definition and of the properties of reflexivity, anti-symmetry, and transitivity of partial orderings, we have:

(2.2.2)   REMARK: If R is a partial ordering and x is in the field of R, then $sup_R\{x\}$ exists; indeed, $sup_R\{x\} = x$.

(2.2.3)   REMARK: If $sup_R S$ and $sup_R T$ both exist and if for every x in S there is a y in T such that xRy, then $sup_R S$ bears R to $sup_R T$.

INDIVIDUALS 33

Another concept frequently employed in Goodman's discussion of individuals is that of an 'atom'. Again, the notion of an atom should be defined at least for every relation R which might qualify as a part-whole relation. Furthermore, given such a relation and an object entering the relation, that object might be regarded as atomic in one context and as composite in another (the intended sense of 'context' will be clarified later). For this reason, we also want to relativize the notion of an atom to a set S which represents, so to speak, the setting in which the object is to be atomic. The desired notion is closely similar to that of an R-minimal element (1.3.1.34) and is defined as follows:

(2.2.4) DEFINITION: *x is R-least in S* [or, x is an R-least element of S] if and only if $x \in S$ and there is no y such that $y \in S$, yRx and $y \neq x$.

Omitting reference to S, we say simply that x is an R-least element if x is an R-least element relative to the whole field of entities which can enter the relation R:

(2.2.5) DEFINITION: *x is R-least* just in case x is R-least in the field of R.

Next, we characterize the central notion of a part-whole relation. We shall first state and then explain the definition:

(2.2.6) DEFINITION: **R** is a *part-whole relation* if and only if the following conditions are satisfied:
(1) **R** is a partial ordering,
(2) 0 is not a member of the field of **R**, and
(3) there exists a set A meeting the following requirements:
   (a) for every non-empty subset S of A and for all x in A, if x bears **R** to $\sup_R S$, then x is in S,
   (b) the field of **R** = the set of all items x such that for some non-empty subset S of A, $x = \sup_R S$, and
   (c) A is infinite.

Throughout the discussion of this definition, it is assumed that **R** is a part-whole relation and that A is a set satisfying the conditions (3a), (3b), and (3c).

The pre-systematic notion of a part-whole relation seems to allow of two possibilities: that everything is part of itself, or that nothing is part of itself. Technically, it is slightly more convenient to embrace the first alternative; and once we do, it is plausible to require that only identical objects shall be part of one another, and that parts of the parts of an object shall themselves be parts of that object. This is the intuitive content of (1).

The condition (2), that the empty set shall not be in the field of a part-whole relation, has been inserted to insure that all terms which are defined by descriptive phrases have their intended meaning when they are referring to entities which enter a part-whole relation. The following remark will illustrate this point:

(2.2.7)   REMARK: For every non-empty subset S of A, $\sup_\mathbf{R} S$ exists.

For, if S is any non-empty subset of A, $\sup_\mathbf{R} S$ is in the field of $\mathbf{R}$ (by 3b) and differs from 0 (by 2). Since, by the convention (1.3.1.1), improper descriptions equal 0, the description defining '$\sup_\mathbf{R} S$' must be a proper one. Hence, $\sup_\mathbf{R} S$ exists.

Of course, we could have replaced the condition (2) of the definition by (2.2.7) itself. But it is more convenient to lay down a condition which assures, once and for all, the properness of all descriptions yet to be introduced in the relevant contexts.

The condition (3a) provides for a set A of objects which have the following characteristics:

To begin with, no element of A is a proper part (in the sense of $\mathbf{R}$) of any other element of A. Indeed, we can say:

(2.2.8)   REMARK: Every element of A is $\mathbf{R}$-least.

For, assuming that $x \in A$, $\{x\} \subseteq A$. By (2.2.2), $x = \sup_\mathbf{R}\{x\}$. Hence, for any y, if $y\mathbf{R}x$ then y bears $\mathbf{R}$ to $\sup_\mathbf{R}\{x\}$ and thus, by (3a), $y = x$.

This justifies our calling A a set of atoms. In addition, it is never possible that an atom in A is part of the sum of two other atoms in A:

(2.2.9)   REMARK: If x, y, and z are in A and $x \neq y$ and $x \neq z$, then it is not the case that x bears $\mathbf{R}$ to $\sup_\mathbf{R}\{y, z\}$.

If an atom x could be part of the sum of two other atoms y and z, then x would 'overlap' and fail to be 'discrete from' y and z. Quite generally, the condition (3a) provides that all atoms in A are 'discrete' in the sense that no atom can be a proper part of any sum of numerous atoms, all different from the given one. In addition to meeting the intuitive requirement that atoms should be discrete, the condition (3a) will also turn out to be formally necessary in subsequent proofs.

The condition (3b) of (2.2.6) asserts that all and only those objects can enter the part-whole relation **R** which are suprema, relative to **R**, of non-empty subsets of the class A of atoms. It embodies Goodman's condition that, relative to a given part-whole relation, just those entities can qualify as individuals which are either atoms or 'generated' from atoms. In this connection, we take it that to be 'generated' from atoms is to be a non-empty sum of atoms.

(3b) has the following useful corollary:

(2.2.10) LEMMA: If S and T are non-empty subsets of A and $\sup_\mathbf{R} S = \sup_\mathbf{R} T$, then $S = T$.

For, assuming the hypothesis of the lemma, every element x of S bears R to $\sup_\mathbf{R} S$ (by 2.2.7 and definitions); hence x bears R to $\sup_R T$ and (by 3a of 2.2.6) $x \in T$. Thus, $S \subseteq T$. Similarly, $T \subseteq S$. Hence, $S = T$.

The condition (3b) also has the consequence that all atoms are members of A:

(2.2.11) REMARK: If x is **R**-least then $x \in A$.

For, if x is **R**-least, then x is in the field of **R**. Hence, for some non-empty subset S of A, $x = \sup_\mathbf{R} S$. Let $y \in S$. yRx (by 2.2.7 and definitions). Since x is **R**-least, $y = x$. But $y \in A$; so, $x \in A$.

Due to the remarks (2.2.8) and (2.2.11), we are justified in calling A *the* set of atoms of the given part-whole relation.

The formal justification of condition (3c), namely, that there shall be infinitely many atoms, will appear later. Its intuitive plausibility will be discussed in the next section.

We note, in passing, that the identity relation (restricted to any set) cannot be regarded as a part-whole relation (due to (3c) and (3b) of 2.2.6). Also, the field of a part-whole relation cannot comprise an element which is part of everything. In this sense, there can be no null-element relative to a part-whole relation.

Every part-whole relation can be represented by the inclusion relation restricted to some infinite set not comprising the empty set. In stating this result, we employ the notion of isomorphic relations (defined in

1.3.1.53), and those of a power set (1.3.1.10), of set theoretic difference (1.3.1.9), and of inclusion restricted to a set (1.3.1.37):

(2.2.12) REPRESENTATION THEOREM: **R** is a part-whole relation if and only if for some infinite set S, **R** is isomorphic with inclusion restricted to the power set of S minus $\{0\}$.

Assuming that **R** is a part-whole relation, let S be the set of all **R**-least elements. The isomorphism is given by that function f whose domain is the field of **R** and which is such that, for all x in its domain, $f(x) =$ the union of the set of all $\{y\}$ where y is an **R**-least element and y**R**x. It is not hard to verify that f is one-to-one, that its range is the power set of S minus $\{0\}$, and that inclusion restricted to its range meets the conditions of (2.2.6) if the set A of **R**-least elements is taken to be the set of all singletons of members of S. Conversely, if S is any infinite set, inclusion restricted to $S \sim \{0\}$ is a part-whole relation having the singletons of members of S for its atoms. If **R** is isomorphic to this relation, it follows from the properties of isomorphism that **R** is a part-whole relation.

Since all part-whole relations are structurally similar to the inclusion relation among infinitely many non-empty sums of singletons, their properties can be studied in terms of the more familiar inclusion relation. Due to a similar representation theorem pertaining to Boolean algebras, part-whole relations turn out to be, essentially, just complete and atomic Boolean algebras with infinitely many elements, but with the null-element removed.[9] In this fashion, part-whole relations, although they were not originally characterized with this end in view, are now seen to have well-known algebraic properties.

2.3. ATOMISTIC UNIVERSES OF INDIVIDUALS

The discussion in Section 2.1 suggests that Goodman's conception of an individual finds expression in two principles: (1) a principle of individuation akin to (2.1.1), and (2) a principle of summation appropriate to the relation between fragments and wholes. Such principles do not permit us state, without circularity, what an individual *is*, but only what characteristic relations obtain between entities of the kind 'individual'. The basic notion appears to be that of a totality of individuals or, as we shall say, of a universe of individuals, and not that of an individual. Thus, although we cannot differentiate directly between individuals and non-individuals, a totality of individuals is distinguished from other totalities by the fact that any two entities in it differ just in case they have different atomic parts.

Given the concept of a part-whole relation, as defined in the preceding section, we shall seek to formulate appropriate principles of individuation and of summation in the form of conditions imposed on sets of entities which can enter such a relation. The subsets of the field of a part-whole relation which meet these conditions of individuation and summation, will be called 'universes of individuals'. An individual will then be indirectly characterized as any member of such a universe of individuals.

Suppose that **R** is a part-whole relation and that **U** is a subset of the field of **R**. Then the following principles of individuation might appear to be suitable conditions on **U**, if **U** is to be a universe of individuals in the intended sense:

(2.3.1)   For all x and y in **U**: $x=y$ if and only if for all z in **U**, $z\mathbf{R}x$ just in case $z\mathbf{R}y$.

Informally: individuals (in **U**) are identical if and only if they have the same **R**-parts in **U**.

(2.3.2)   For all x and y in **U**: $x=y$ if and only if for all z, if z is **R**-least then $z\mathbf{R}x$ just in case $z\mathbf{R}y$.

Informally: individuals (in **U**) are identical if and only if they have the same atoms, relative to the whole field of **R**, for parts.

(2.3.3)   For all x and y in **U**: $x=y$ if and only if for all z, if z is **R**-least in **U**, then $z\mathbf{R}x$ just in case $z\mathbf{R}y$.

Informally: individuals (in **U**) are identical if and only if they have the same atoms, relative to **U**, for parts.

In order to select the most plausible of these principles, the following may be noted:

Principle (2.3.1) imposes no restriction: every subset **U** of the field of **R** satisfies (2.3.1) by virtue of the reflexivity and anti-symmetry of R. Also, (2.3.1), lacking all reference to atoms, is not a plausible formulation of that principle of individuation which was discussed in Section 2.1.

Principle (2.3.2) also imposes no restriction in that it is provable.

For, assume that x and y are in **U** and that for every **R**-least element z, $z\mathbf{R}x$ just in case $z\mathbf{R}y$. Let A be the set of **R**-least elements. Since **U** is a subset of the field of **R** and by 2.2.6 (3b), there exist non-empty subsets X and Y of A such that $x = \sup_\mathbf{R} X$ and $y = \sup_\mathbf{R} Y$. By (2.2.7), these suprema exist. For every element z of X, $z\mathbf{R}x$ and hence $z\mathbf{R}y$. By definition, $\sup_\mathbf{R} X \, (=x)$ bears $\mathbf{R}$ to y. Similarly, $y\mathbf{R}x$. By antisymmetry, $x = y$.

Principle (2.3.3), on the other hand, is not satisfied in every subset of the field of $\mathbf{R}$. To see this, consider an example (which will be used repeatedly):

(2.3.4)   EXAMPLE: Let A_E be an infinite set of singletons; let $F_E =$ the set of all unions of non-empty subsets of A_E; and let $R_E =$ the inclusion relation restricted to F_E. Clearly, R_E is a part-whole relation. Assume that A, B, C, and D are distinct objects whose singletons are in A_E.

Under these assumptions, let $U_E = \{\{A\}, \{A, B\}\}$. U_E is a subset of the field F_E of R_E which, like every such subset, satisfies the conditions (2.3.1) and (2.3.2). But U_E fails to meet the condition (2.3.3). For $\{A\}$ is the only R_E-least element of U_E. Hence, $\{A\}$ and $\{A, B\}$ have the same atomic parts relative to U_E. Yet, $\{A\} \neq \{A, B\}$.

In order to decide whether the restriction afforded by (2.3.3) is a plausible one, we must consult our intuition regarding the role of universes of individuals within the fields of part-whole relations. According to our favorite informal interpretation of these set-theoretic entities, the field of a part-whole relation is regarded as the set of all *possible* objects which can be parts or wholes in the sense of that relation.[10] By contrast, the elements of a particular universe of individuals are regarded as those individuals which happen to be *actualized* in that universe. To provide a suggestive example: suppose that we conceive of an infinite class of items all of which satisfy a physicist's description of an atom. Let a part-whole relation be conceived between these atoms and all possible composites of the atoms. Any selection of these possible atoms or composites might be actualized in some universe which is a 'universe of individuals' if for every composite object which is actualized in it a sufficient variety of parts are also actualized, so that different actual composites have in the universe different actual parts. It is logically possible that the simplest physical objects which happen to be actualized in such a universe are molecules, while all proper parts of molecules remain unactualized possibles. In such a state of affairs it would remain logically possible that molecules have proper parts, and we would know what it means to say that they have such parts; but the splitting of molecules would not, in this universe, be a physical actuality.

A universe of individuals may be included in the fields of two or more part-whole relations. For instance, a universe comprising just the set $\{A, B\}$ may have the 'unactualized' parts A and B relative to chains of membership, but the 'unactualized' parts $\{A\}$ and $\{B\}$ relative to inclusion. We regard the members of such a universe as wholes relative to as many different principles of division as there are part-whole relations which have the universe as a common subset of their fields. The items which enter two or more part-whole relations are conceived as possible entities whose entering into the various relations is 'compossible'. Given any two distinct part-whole relations, their fields will differ, and hence there will be entities which cannot possibly be construed as parts or wholes with respect to both of these relations at once.

A nominalist of the variety considered in Section 2.1 is presumed to hold that distinct actual individuals are composed of distinct actual atoms – which, given the present informal interpretation, is the content of principle (2.3.3). It seems contrary to the spirit of nominalism to regard hypothetical entities as individuals. For this reason, a nominalist is not likely to hold that distinct actual individuals must be composed of distinct possible atoms – an informal interpretation of principle (2.3.2). We conclude by regarding (2.3.3) as the most plausible of the proposed principles of individuation.

A side-remark completing the discussion given in the preceding section of part-whole relations: The condition (3c) of the definition (2.2.6) required that the set of atoms of a part-whole relation shall be infinite. Given our present informal interpretation of such relations, the atoms in question are construed as possible, rather than as actual objects. And it does not seem counter-intuitive to require that there shall be infinitely many possible entities. On the other hand, since universes of individuals are conceived as comprising actual individuals, we shall refrain from imposing a condition on such universes which would imply that every universe of actual things is infinite. In this respect we follow Goodman[11] by remaining neutral between the positions of finitism and non-finitism.

Let us turn next to a comparison of some principles of summation. Suppose that **R** is a part-whole relation and **U** is a subset of the field of **R**. Then the following principles might seem appropriate, if **U** is to qualify as a universe of individuals:

(2.3.5)   For every x in **U** there exists a subset S of the field of **R** such that $x = \sup_\mathbf{R} S$.

Informally: every individual (in **U**) is a sum of 'possible' **R**-parts (not necessarily occurring in **U**).

(2.3.6)   For every x in **U** there exists a subset S of **U** such that $x = \sup_\mathbf{R} S$.

Informally: every individual (in **U**) is a sum of 'actual' **R**-parts occurring in **U**.

(2.3.7)   For every x in **U** there exists a set S of **R**-least elements such that $x = \sup_\mathbf{R} S$.

Informally: every individual (in **U**) is a sum of elements which are atomic relative to the entire field of **R**.

(2.3.8)   For every x in **U** there exists a set S of items which are **R**-least in **U** such that $x = \sup_\mathbf{R} S$.

Informally: every individual (in **U**) is a sum of elements which are atomic relative to **U**.

(2.3.9)   For every non-empty subset S of **U**, $\sup_\mathbf{R} S$ is in **U**.

Informally: the proper sum of any individuals (in **U**) is always again an individual (in **U**).

There is a striking difference between the first four principles and the last one. The former might be called 'division principles': they assert that each individual can be divided into parts of a certain sort. The latter, (2.3.9), is a closure principle: it asserts that the universe of individuals is closed under sums of its non-empty subsets.

Goodman, by his own admission[12], has been most frequently criticised for adopting the principle that the sum of any individuals is again an individual; that is, for (2.3.9) whose syntactical analog is embodied in the Leonard-Goodman calculus of individuals. Indeed, this principle has counter-intuitive instances. Consider, e.g., a part-whole relation between physical objects which is so understood that each part of a whole is in close spatial proximity of some other part (discrete from the first) of the same whole. This relation, which excludes composites of widely scattered fragments, seems to be the most natural relation of part to whole for entities which would be called 'physical objects' for the purposes of ordinary (as opposed to scientific) discourse. Regard the actual universe of physical objects as a universe of individuals. Then, e.g., the sum of two distant stars cannot be regarded as another individual. Yet Goodman seems to hold that the sum of any two individuals relative to any 'generating relation' is always again an individual. Hence, if Goodman intended to allow for 'generating' or part-whole relations of which the one just mentioned is an example, then his principle of summation is clearly too strong. On the other hand, it could be that Goodman's conception of a part-whole relation deviates from the customary one to the extent that it is analytic of his notion that all sums of individuals exist relative to any relation which he would be prepared to call a 'part-whole' relation. In that event, we would depart from Goodman's conception by admitting other relations which qualify intuitively as part-whole relations but fail to generate actual sums of arbitrary individuals. Accordingly, we refrain from imposing the principle of summation (2.3.9) on all universes of individuals. A treatment shall however be given, in subsequent sections, of the special species of universes which satisfy a principle of this sort.

Having rejected (2.3.9) as a general condition on universes of individuals, we turn to an evaluation of the remaining principles.

We note that the principles (2.3.5)–(2.3.7) impose no restriction whatever, since they are easily provable (by appeals to (2.2.2) and to the condition (3b) of (2.2.6)).

The principle (2.3.8) of summation implies the principle (2.3.3) of individuation, as indicated by the theorem (2.3.13) yet to be stated. Since that principle of individuation has seemed to us the most plausible candidate among rival principles, at least part of the content of (2.3.8) is equally plausible.

The principle of individuation (2.3.3), however, does not imply the principle of summation (2.3.8). To see this, consider again the part-whole relation R_E given in the example (2.3.4). Let a new universe of individuals $U_E = \{\{A\}, \{B\}, \{A, B, C\}\}$. The elements $\{A\}$ and $\{B\}$ are the only atoms, relative to inclusion, of U_E, and there is no set S of atoms of U_E whose union is $\{A, B, C\}$. But inspection shows that the principle of individuation (2.3.3) holds in U_E.

Another principle is helpful in evaluating (2.3.8). It seems plausible that an individual in a given universe should be a least whole composed of those atomic parts which occur in that universe. This can be informally expressed as follows:

(2.3.10) PRINCIPLE: If x is an element of a universe of individuals and y is in the field of the appropriate part-whole relation, and if every atom (relative to the universe) which is part of x is also part of y, then x itself is part of y.

This principle is formulated precisely in lemma (2.3.12) below, whose proof makes it clear that the principle is equivalent with the principle of summation (2.3.8). That (2.3.10) is not implied by the principle of individuation (2.3.3) can be directly verified by the last example. Let $x=\{A, B, C\}$ and let $y=\{A, B\}$. y is in the field of the part-whole relation R_E, and every atom in U_E which is included in x is included in y. Yet x is not included in y.

We have selected the principle of summation (2.3.8), and by implication the principle of individuation (2.3.3), as the appropriate conditions on universes of individuals which approximate Goodman's account of individuals (in 1956). Since these principles imply that there exist atoms in each universe, and in distinction from other universes to be considered later, we call such universes 'atomistic':

(2.3.11) DEFINITION: **U** *is an atomistic universe of individuals for* **R** if and only if **R** is a part-whole relation, **U** is included in the field of **R**, and (2.3.8) for every x in **U** there exists a set S such that all members of S are **R**-least in **U** and $x = \sup_\mathbf{R} S$.

For brevity, we shall often speak of a 'universe of individuals' or of a 'universe' in contexts where no misunderstanding seems possible.

The following lemma asserts the equivalence between the principle of summation (2.3.8) and the condition informally expressed by Principle (2.3.10):

(2.3.12) LEMMA: Suppose that **R** is a part-whole relation and that **U** is included in the field of **R**. Then **U** is an atomistic universe of individuals for **R** if and only if the following condition (C) is satisfied: for every x in **U** and for every

INDIVIDUALS   43

y in the field of **R**: if every **R**-least element of **U** which bears **R** to x bears **R** to y, then x bears **R** to y.

PROOF: Assume the hypothesis of the lemma.

SUFFICIENCY: Assume (1) that **U** is an atomistic universe for **R**, (2) that $x \in $**U**, (3) that y is in the field of **R**, and (4) that every **R**-least element of **U** which bears **R** to x bears **R** to y. By (1), (2), and the definition, there exists a set S such that all members of S are **R**-least in **U** and $x = \sup_\mathbf{R} S$. Clearly, $\sup_\mathbf{R} S$ exists; for if it did not, it would be the case that $x = 0$, and 0 is in the field of **R** which is impossible. Show (5) that every member of S bears **R** to y. Assume that $z \in S$, so that z is **R**-least in **U**. Since $\sup_\mathbf{R} S$ exists, z bears **R** to $\sup_\mathbf{R} S$ and hence to x. By (4) we infer zRy, which establishes (5). By the definition of a supremum and (5), $\sup_\mathbf{R} S$ bears **R** to y, and therefore so does x.

NECESSITY: Assume the condition (C) of the lemma. We want to show that for every x in **U** there is a set S such that all members of S are **R**-least in **U** and $x = \sup_\mathbf{R} S$. Let $S =$ the set of all z such that z is **R**-least in **U** and zRx. We have to show that $\sup_\mathbf{R} S$ exists. To that end, let $A =$ the set of **R**-least elements, and let $T =$ the set of all $a \in A$ such that aRx.

(1) Show: for all z in S, z bears **R** to $\sup_\mathbf{R} T$. Assume that $z \in S$, so that z is **R**-least in **U** and zRx. z is in the field of **R**, and therefore there is a non-empty subset Z of A such that $z = \sup_\mathbf{R} Z$. We establish (A) that every element of Z bears **R** to $\sup_\mathbf{R} T$. For suppose that $a \in Z$. $\sup_\mathbf{R} Z$ exists and is equal to z. Hence, aRz. Since also zRx and by transitivity, aRx. Hence, $a \in T$. Since x is in the field of **R**, clearly, $\sup_\mathbf{R} T$ exists. Thus, (A) follows. By (A) and the fact that $\sup_\mathbf{R} Z$ exists, $\sup_\mathbf{R} Z$ bears **R** to $\sup_\mathbf{R} T$. Hence, z bears **R** to $\sup_\mathbf{R} T$.

(2) Show: for all t, if every member of S bears **R** to t, then $\sup_\mathbf{R} T$ bears **R** to t. Assume that every member of S bears **R** to t. Show (A): every member of T bears **R** to t. Suppose $a \in T$, so that $a \in A$ and aRx. Show (B): every **R**-least element of **U** which bears **R** to x bears **R** to t. Assume that u is **R**-least in **U** and uRx. Thus, $u \in S$. Hence, by assumption, uRt, establishing (B). From the condition (C) and from (B) it follows that xRt. By transitivity, aRt. So, (A) has been shown. Since $\sup_\mathbf{R} T$ exists and by (A), $\sup_\mathbf{R} T$ bears **R** to t.

By (1) and (2) and the definition of suprema, $\sup_\mathbf{R} S$ exists.

Further, since every element of S bears **R** to x, $\sup_\mathbf{R} S$ bears **R** to x. Also, $\sup_\mathbf{R} S$, being identical with $\sup_\mathbf{R} T$, is in the field of **R** and every **R**-least element of **U** which bears **R** to x also bears **R** to $\sup_\mathbf{R} S$. By the condition (C), x bears **R** to $\sup_\mathbf{R} S$. By antisymmetry, $x = \sup_\mathbf{R} S$. This completes the proof.

It has been claimed that every universe of individuals, due to the condition (2.3.8), also satisfies the principle of individuation (2.3.3). This assertion is expressed in the following theorem which is an immediate consequence of Lemma (2.3.12):

(2.3.13)   THEOREM: Suppose that **U** is an atomistic universe of individuals for **R**. Then (2.3.3), for all x and y in **U**: $x = y$ if and only if for all z, if z is **R**-least in **U**, then zRx just in case zRy.

In this fashion, we hope to have captured the essential ideas expressed in Goodman's semi-formal account of individuals. We shall next turn to his formal contributions.

2.4. THE LEONARD-GOODMAN CALCULUS OF INDIVIDUALS LGCI

Throughout our discussion of individuals so far, we have taken for granted the deductive apparatus of set theory and the existence of various sets. However, at least some of these suppositions are rejected by contemporary nominalists. For this reason it will be our aim to specify formal deductive systems which are acceptable to a nominalist and which represent that portion of the theory of individuals which can be expressed in such systems. More specifically, we seek to formulate an axiomatic system whose theorems, under a natural interpretation, are just those sentences of the system which are true in all universes of individuals. Systems of this sort are called 'calculi of individuals'. The theorems of such a calculus are regarded as necessarily true of individuals just as, in perhaps more familiar cases, the theorems of a calculus of classes or of relations are regarded as necessarily true of classes or of relations.

Leonard and Goodman (1940) have presented such a calculus of individuals[13] without, however, demonstrating either its soundness or its completeness relative to a precise semantical interpretation. For the convenience of the reader, we reconstruct here a formal system similar to their original calculus of individuals. We depart in our reconstruction from the earlier calculus in only one respect which might be taken to have philosophical significance:

The Leonard-Goodman calculus (in 1940) is an extension of the *Principia mathematica*. But a nominalist of Goodman's persuasion could hardly endorse the theory of classes which is contained in this underlying logic. Furthermore, one of the postulates peculiar to the calculus of individuals in question asserts the existence of a sum or fusion of all individuals in a given class whenever that class is non-empty. We shall avoid this reference to classes by substituting an axiom scheme comprehending all specifiable conditions of sum-formation.[14]

Our reconstruction of the Leonard-Goodman Calculus of Individuals (for short, LGCI) conforms to the description of formal systems given in Section 1.3.2.

INDIVIDUALS

The *vocabulary* of LGCI comprises (apart from a denumerable set of variables, the parentheses, sentential connectives, and quantifiers) two *primitive constants*: the two-place non-schematic predicate '∘', and the two-place non-schematic predicate '='; and three *defined constants*: the two-place predicates '⩽' and '<', and the one-place predicate '*At*'. There are no other predicates or operation symbols.

To be a *term* of LGCI is to be a variable.[15]

Given any terms ζ and η, the following are *atomic formulas* of LGCI:

$\zeta \circ \eta$ [read: ζ overlaps η],

$\zeta = \eta$ [read: ζ is identical with η],

$\zeta \leqslant \eta$ [read: ζ is a part of η],

$\zeta < \eta$ [read: ζ is a proper part of η], and

At ζ [read: ζ is an atom].

All atomic formulas are *formulas* of LGCI; and whenever ϕ and ψ are formulas and α is a variable, the concatenates $\ulcorner \sim \phi \urcorner$, $\ulcorner (\phi \to \psi) \urcorner$, $\ulcorner (\phi \& \psi) \urcorner$, $\ulcorner (\phi \vee \psi) \urcorner$, $\ulcorner (\phi \leftrightarrow \psi) \urcorner$, $\ulcorner (\forall \alpha) \phi \urcorner$, and $\ulcorner (\exists \alpha) \phi \urcorner$ are again formulas of LGCI. Pairs of outermost parentheses and of ones occurring in repeated conjunctions or repeated disjunctions may be omitted. Pairs of brackets may replace corresponding pairs of parentheses.

For all terms ζ and η, for all variables α, and for all formulas ϕ, ϕ', and ψ such that ϕ' results from ϕ by replacing one or more free occurrences of ζ by free occurrences of η, the following are taken as the *logical axioms* of LGCI:

(L1) all tautologies,

(L2) $\ulcorner (\forall \alpha)(\phi \to \psi) \to [(\forall \alpha)\phi \to (\forall \alpha)\psi] \urcorner$,

(L3) $\ulcorner (\exists \alpha) \phi \leftrightarrow \sim (\forall \alpha) \sim \phi \urcorner$, and

(L4) $\ulcorner (\zeta = \eta \& \phi) \to \phi' \urcorner$.

Further axioms, whose content is chiefly logical, will contain the special notation of LGCI and are therefore listed among the proper axioms of LGCI.

For all terms ζ and η and for all variables α not occurring in either ζ or η, the following are *definitional axioms* of LGCI:

(D1) $\ulcorner \zeta \leqslant \eta \leftrightarrow [\zeta \circ \zeta \& (\forall \alpha)(\alpha \circ \zeta \to \alpha \circ \eta)] \urcorner$.

In words: ζ is a part of η just in case ζ overlaps itself (or, ζ exists) and every individual which overlaps ζ overlaps η.

(D2) $\ulcorner \zeta < \eta \leftrightarrow (\zeta \leqslant \eta \ \& \sim \eta \leqslant \zeta) \urcorner$.

In words: ζ is a proper part of η just in case ζ is a part of η, but η is not a part of ζ.

(D3) $\ulcorner At\,\zeta \leftrightarrow [\zeta \circ \zeta \ \& \sim (\exists \alpha)(\alpha < \zeta)] \urcorner$.

In words: ζ is an atom just in case ζ overlaps itself (or, ζ exists) and has no proper parts.

For all terms ζ and η, for all distinct variables α, β, and γ such that β does not occur in ζ or η, and for all formulas ϕ such that β and γ do not occur in ϕ, the following are *proper axioms* of LGCI:

(A1) $\ulcorner (\zeta \circ \zeta \ \& \ (\forall \alpha)\phi) \to \phi^{\alpha}_{\zeta} \urcorner$.

In words: If ζ overlaps itself (or, ζ exists) and if everything satisfies the condition ϕ, then so does ζ. As indicated in (1.3.2.3), the notation 'ϕ^{α}_{ζ}' refers to the proper substitution of the term ζ for the variable α throughout ϕ. (A1) expresses that portion of the law of Universal Specification which can be expected to hold under an implicit interpretation which allows for the empty universe of discourse.

(A2) $\ulcorner (\forall \alpha)\, \alpha \circ \alpha \urcorner$.

In words: Every (actual) individual overlaps itself; or, every (actual) individual exists.

(A3) $\ulcorner \zeta = \zeta \to \zeta \circ \zeta \urcorner$.

In words: If ζ is self-identical, then ζ overlaps itself (or, ζ exists). The converse of (A3) follows essentially from (A4) below and from (D1).

The logical axioms (L1)–(L4), together with the axioms (A1), (A2), and the result of replacing the conditional in (A3) by a biconditional, constitute a theory closely akin to a system of logic with identity whose completeness has been demonstrated elsewhere[16] and in whose interpretation allowance is made for the empty universe of discourse and, in general, for denotationless terms of arbitrary complexity. The calculus of individuals permits that we translate sentences of the form $\ulcorner \zeta$ exists$\urcorner$ or express that ζ is a denoting term by the formula $\ulcorner \zeta \circ \zeta \urcorner$ instead of using

INDIVIDUALS

⌜$\zeta = \zeta$⌝ or, if identity were not available, by a special predicate expressing existence.

(A4)   ⌜$(\zeta \leqslant \eta \ \& \ \eta \leqslant \zeta) \rightarrow \zeta = \eta$⌝.

In words: Individuals which are part of one another are identical.

(A5)   ⌜$\zeta \circ \eta \leftrightarrow (\exists \beta)(\beta \leqslant \zeta \ \& \ \beta \leqslant \eta)$⌝.

In words: Individuals overlap just in case they have a common part.

(AS6)   ⌜$(\exists \alpha) \phi \rightarrow (\exists \gamma)(\forall \beta)[\beta \circ \gamma \leftrightarrow (\exists \alpha)(\phi \ \& \ \beta \circ \alpha)]$⌝.

In words: If something satisfies the condition ϕ, then there exists an individual which overlaps all and only those things which overlap something satisfying ϕ. Or, informally, if something satisfies the condition ϕ, then the sum of all individuals satisfying ϕ exists.

This completes the list of axioms. The *inference rules* are Modus Ponens and Universal Generalization on the consequent of conditionals (see 1.3.2.4).

Among the specifically logical features of LGCI we mention here, without proof, only the following: Under the common hypothesis prefacing the definition of proper axioms of LGCI, the following are theorems of LGCI:

(2.4.1)   ⌜$(\zeta \circ \zeta \ \& \ \phi_\zeta^\alpha) \rightarrow (\exists \alpha) \phi$⌝.

In words: If ζ overlaps itself (or, if ζ exists) and satisfies the condition ϕ, then something satisfies the condition ϕ (by A1). That is to say, existential generalization is justified provided the term ζ, on which one generalizes, is a denoting term.

(2.4.2)   ⌜$(\forall \alpha)(\phi \rightarrow \psi) \rightarrow (\phi \rightarrow (\forall \alpha) \psi)$⌝,

provided ϕ and ψ are formulas and α is not free in ϕ (by A1, generalization, L2, and A2).

(2.4.3)   ⌜$(\forall \alpha) \sim \alpha \circ \alpha \rightarrow (\forall \alpha) \phi$⌝.

In words: If all individuals fail to overlap themselves (that is, if there are no individuals whatever), then every (actual) individual satisfies the condition ϕ (by A1, L2, A2). Further, if β is not free in ϕ,

(2.4.4)   ⌜$(\forall \beta) \phi_\beta^\alpha \rightarrow (\forall \alpha) \phi$⌝.

That is to say, rewrite of bound variables holds (by A1, L2, A2). The Deduction Theorem is provable of LGCI.

In previous discussions a principle of individuation has been adopted, according to which individuals (actualized in a given universe of discourse) are identical if they have the same atoms for parts (the parts being atoms in the same universe). In LGCI the syntactical counterpart of this principle can be expressed as follows:

(2.4.5) $\ulcorner (\forall x)(\forall y) [(x \circ x \ \& \ y \circ y \ \&$
$(\forall z)[Atz \to (z \leqslant x \leftrightarrow z \leqslant y)]) \to x = y]\urcorner.$

In words: Individuals which exist and which have the same atoms for parts are identical.

Since Goodman, at least in later discussions (in 1956), regards such a principle as central to the very notion of an individual, one would expect (2.4.5) (or a definitional equivalent) to be a theorem of that calculus which was intended to treat specifically of individuals. Surprisingly, this is not the case.

To see this, we note first a few elementary consequences of the axioms: The relation of being a part, '$\leqslant$', is antisymmetric and transitive, but reflexive only in the sense that $\ulcorner x \circ x \to x \leqslant x \urcorner$ holds. The relation of being a proper part, '$<$', is irreflexive, asymmetric and transitive. Overlapping, '$\circ$', is symmetric. We observe that the principle of individuation (2.4.5) is equivalent, within LGCI, to

(2.4.6) $\ulcorner (\forall x) [x \circ x \to (\exists y)(Aty \ \& \ y \leqslant x)]\urcorner.$

In words: Every individual which overlaps itself has an atomic part.

An informal outline of the proof proceeds as follows:
To show that (2.4.5) implies (2.4.6), assume (2.4.5) and also that $x \circ x$. Suppose, contrary to (2.4.6), that x has no atomic part. Then x itself is not an atom. By definition (D3), and since x overlaps itself, there is some individual y which is a proper part of x. By the definitions, y is part of x and y overlaps itself. y also has no atomic parts. For, if z were such a part, by transitivity, z would be an atomic part of x, z (by definition) would overlap itself, and hence x would have an atomic part. Thus, the antecedent in (2.4.5) is (in part vacuously) true of x and y. Therefore, $x = y$ contradicting the assertion that y is a proper part of x.

In order to show that (2.4.6) implies (2.4.5), assume the former and suppose, contrary to (2.4.5), that both x and y overlap themselves, that x and y have the same atomic parts, but that $x \neq y$. By (A4), one of x and y fails to be part of the other. Without loss of generality, assume that x fails to be part of y. By (D1) and since x overlaps itself, there exists an individual z such that z overlaps x but not y. By (A5),

INDIVIDUALS   49

some individual u is part of both z and x. u overlaps itself. Hence, by (2.4.6) and (A1), there exists an atomic part v of u. By definition, v overlaps itself and, by transitivity, v is part of x. Since x and y have the same atomic parts, v is also part of y. Since v is part of u and u is part of z, v is part of z. v, being part of both z and y and overlapping itself, we conclude that there exists a common part of both z and y. Hence, by (A5), z overlaps y, contrary to our earlier observation.

It is easy to see that (2.4.6), in turn, is equivalent to:

(2.4.7) $\ulcorner (\forall x)(\exists y)(Aty \ \& \ y \leqslant x) \urcorner$.

In words: Every (actual) individual has an atomic part.

For, if ϕ is any formula, $\ulcorner (\forall x)(x \circ x \to \phi) \to (x \circ x \to \phi) \urcorner$ is tautologically equivalent to an instance of (A1). By generalization on the consequent and distribution in accord with (L2), $\ulcorner (\forall x)(x \circ x \to \phi) \to [(\forall x) x \circ x \to (\forall x)\phi] \urcorner$ is a theorem; and hence so is, by (A2), $\ulcorner (\forall x)(x \circ x \to \phi) \to (\forall x)\phi \urcorner$. That (2.4.6) implies (2.4.7) is an instance of this theorem. The converse implication is obvious.

An interpretation of LGCI proving the independence of (2.4.7) relative to all axioms of LGCI is afforded by a complete atomless system of Boolean algebra upon removal of the null element and by interpreting '$\leqslant$' by the partial ordering of such an algebra.[17]

Since (2.4.5), our formulation of the principle of individuation proposed by Goodman, does not hold in LGCI (the reconstructed Leonard-Goodman system), this system does not appear to be a calculus specifically of individuals in so far as that notion was clarified by Goodman (1956). This fact calls in question the intuitive soundness of the Leonard-Goodman system or shows, at least, a disparity between Goodman's formal account of individuals and his informal account offered more recently.

Instead of LGCI, we shall formalize in the following sections three systems in all of which the principle of individuation (2.4.5) appears as a theorem, and each of which, in this sense, has some claim to being a calculus appropriate to entities in atomistic universes of individuals. Each of these systems will be interpreted, and each is shown to be both sound and complete relative to the given interpretation. The assertions of the first of these systems, CII, shall be true in all atomistic universes of individuals (as characterized in Section 2.3) and relative to all part-whole relations (as defined in Section 2.2). For this reason, we regard this theory as a most comprehensive calculus appropriate specifically to individuals in the sense of Goodman (1956). The second calculus of individuals, CIII, will hold in universes of individuals which are closed, roughly speaking, under non-empty finite intersection and under finite union. Thus, CI II will be regarded as a theory appropriate to individuals satisfying the following conditions: (1) all atoms (relative to the universe) are discrete, and (2) the sum of any two individuals (and hence of any finite number of individuals) is in turn an individual. The third system, CI III, will hold in all atomistic universes of individuals which are closed, roughly, under non-empty intersection and under union of non-empty specifiable subsets of the universe. Thus, CI III will be regarded as a theory appropriate to individuals which satisfy condition (1) mentioned above and which meet in addition the requirement that the sum of any number of individuals – finite or infinite – is always in turn an individual.

These projects leave the question unanswered whether there is a clear notion of an individual which differs from the one so far discussed, and which might serve to interpret the Leonard-Goodman system LGCI. We shall explore this possibility in Section 2.9.

2.5. THE GENERAL CALCULUS CII OF ATOMISTIC INDIVIDUALS[18]

As we have indicated, CII is to be an interpreted formal system whose theorems are true in all atomistic universes of individuals in the sense made precise in previous sections. The characterization of CII follows the pattern described in Section 1.3.2.

2.5.1. *The Vocabulary and the Axioms of CII*

The *vocabulary* of CII is that of the reconstructed Leonard-Goodman Calculus LGCI presented in the preceding section. However, in CII, unlike in LGCI, the identity symbols '=' is treated as a defined constant. Thus, the vocabulary of CII comprises (1) one *primitive constant*, the two-place non-schematic predicate '∘', and (2) four *defined constants*, the one-place non-schematic predicate '*At*', and the two-place non-schematic predicates '⩽', '<', and '='. The *terms* and *formulas* of CII are those of LGCI.

The purely *logical axioms* of CII are the well-known and uncontroversial logical axioms of LGCI under omission of (L4) regarding interchange of identicals. Thus, the logical axioms are just (L1), (L2), and (L3) as previously stated.

In reading the following concatenates, assume throughout that ζ and η are terms, α, β, and γ are distinct variables such that β does not occur in either ζ or η, ϕ is a formula, and ϕ_ζ^α is the proper substitution of the term ζ for the variable α throughout ϕ.

The *definitional axioms* of CII are those of LGCI (which we repeat here for easier reference), together with (D4) which introduces the defined identity sign:

(D1) $\qquad \ulcorner \zeta \leqslant \eta \leftrightarrow [\zeta \circ \zeta \,\&\, (\forall \alpha)(\alpha \circ \zeta \rightarrow \alpha \circ \eta)] \urcorner.$

In words: ζ is part of η just in case ζ overlaps itself (or, ζ exists) and every individual which overlaps ζ overlaps η.

(D2) $\qquad \ulcorner \zeta < \eta \leftrightarrow (\zeta \leqslant \eta \,\&\, {\sim}\eta \leqslant \zeta) \urcorner.$

In words: ζ is a proper part of η just in case ζ is a part of η, but η is not a part of ζ.

(D3) $\qquad \ulcorner At\,\zeta \leftrightarrow [\zeta \circ \zeta \,\&\, {\sim}(\exists \alpha)(\alpha < \zeta)] \urcorner.$

In words: ζ is an atom just in case ζ overlaps itself (or, ζ exists) and has no proper parts.

(D4) $\qquad \ulcorner \zeta = \eta \leftrightarrow (\zeta \leqslant \eta \,\&\, \eta \leqslant \zeta) \urcorner.$

In words: Two individuals are identical just in case each is part of the other.

The *proper axioms* of CII are comprehended by the following list:

(Ax1) $\qquad \ulcorner (\zeta \circ \zeta \,\&\, (\forall \alpha)\phi) \rightarrow \phi_\zeta^\alpha \urcorner,$

In words: If ζ overlaps itself (or, ζ exists) and if everything satisfies the condition ϕ, then so does ζ.

(Ax2) $\qquad \ulcorner (\forall \alpha) \alpha \circ \alpha \urcorner.$

In words: Every (actual) individual overlaps itself (or, every (actual) individual exists).

The first two axioms are just the axioms (A1) and (A2), respectively, of LGCI, and have chiefly logical content.

(Ax3) $\qquad \ulcorner \zeta \circ \eta \leftrightarrow (\exists \beta)(At\,\beta \,\&\, \beta \leqslant \zeta \,\&\, \beta \leqslant \eta) \urcorner.$

Informally: Individuals overlap just in case they have some atom as a common part.

NOMINALISTIC SYSTEMS

Before considering the semantical interpretation of CII, we state a number of theorems of this system which may serve to disclose the intuitive content of the axioms. We refer to (1.3.2.8) for a characterization of the notion of a theorem.

2.5.2. *Some Theorems of CII*

The following formulas are theorems of CII for all terms ζ, η, and τ, for all distinct variables α and β not occurring in ζ, η, or τ, and for all formulas ϕ and ψ such that β is not free in ϕ:

(2.5.2.1) $\quad (\zeta \circ \zeta \,\&\, \phi_\zeta^\alpha) \to (\exists \alpha)\, \phi$.

If ζ overlaps itself (or, if ζ exists) and satisfies the condition ϕ, then something satisfies the condition ϕ.

(2.5.2.2) $\quad (\forall \alpha) \sim \alpha \circ \alpha \to (\forall \alpha)\, \phi$.

If no individuals exist, then all individuals satisfy the condition ϕ.

(2.5.2.3) $\quad (\forall \beta)\, \phi_\beta^\alpha \to (\forall \alpha)\, \phi$.

Rewrite of bound variables holds.

These theorems appeared also as (2.4.2)–(2.4.4) in LGCI.

(2.5.2.4) $\quad (\forall \alpha)(\alpha \circ \alpha \to \phi) \leftrightarrow (\forall \alpha)\, \phi$.

Every individual which overlaps itself (or, exists) satisfies the condition ϕ just in case every (actual) individual satisfies that condition. (The proof of this was outlined in connection with 2.4.7.)

(2.5.2.5) $\quad \zeta \circ \eta \to \eta \circ \zeta$.

'∘' is symmetric.

(2.5.2.6) $\quad \zeta \circ \zeta \to \zeta \leqslant \zeta$.

An individual which overlaps itself (or, exists) is part of itself. The 'quasi-reflexivity' of '$\leqslant$'.

(2.5.2.7) $\quad \zeta = \zeta \leftrightarrow \zeta \circ \zeta$.

An individual is identical with itself just in case it overlaps itself (or, exists).

(2.5.2.8) $(\zeta \leqslant \eta \,\&\, \eta \leqslant \tau) \to \zeta \leqslant \tau$.

A part of a part of an individual is itself a part of that individual. '$\leqslant$' is transitive.

(2.5.2.9) $\zeta \leqslant \eta \to \zeta \circ \eta$.

A part of an individual overlaps that individual.

(2.5.2.10) $\sim \zeta < \zeta$.

An individual is not a proper part of itself. '$<$' is irreflexive.

(2.5.2.11) $\zeta < \eta \to \sim \eta < \zeta$.

Individuals are not proper parts of one another. '$<$' is asymmetric.

(2.5.2.12) $(\zeta < \eta \,\&\, \eta < \tau) \to \zeta < \tau$.

A proper part of a proper part of an individual is itself a proper part of that individual. '$<$' is transitive.

(2.5.2.13) $\zeta < \eta \to (\zeta \leqslant \eta \,\&\, \zeta \circ \eta)$.

A proper part of an individual is part of the individual and overlaps it.

(2.5.2.14) $\zeta = \eta \to \eta = \zeta$.

'$=$' is symmetric.

(2.5.2.15) $(\zeta = \eta \,\&\, \eta = \tau) \to \zeta = \tau$.

'$=$' is transitive.

(2.5.2.16) $\zeta < \eta \to \sim \zeta = \eta$.

A proper part of an individual differs from that individual.

(2.5.2.17) $(\zeta \circ \zeta \,\&\, (\forall \alpha)[\alpha \leqslant \zeta \to \alpha \leqslant \eta]) \to \zeta \leqslant \eta$.

If ζ overlaps itself (or, exists), and if every part of ζ is part of η, then ζ is part of η. (By D1 and Ax1, Ax3.)

(2.5.2.18) $\zeta \circ \eta \to \zeta \circ \zeta$.

If ζ overlaps η then ζ overlaps itself (or, ζ exists).

(2.5.2.19) $\zeta = \eta \leftrightarrow (\forall \alpha)(\alpha \leqslant \zeta \leftrightarrow \alpha \leqslant \eta) \mathbin{\&} \zeta \circ \zeta \mathbin{\&} \eta \circ \eta$.

Two individuals are identical just in case they have the same parts and exist.

(2.5.2.20) $\zeta \circ \eta \leftrightarrow (\exists \alpha)(\alpha \leqslant \zeta \mathbin{\&} \alpha \leqslant \eta)$.

Two individuals overlap just in case they have a common part.

(2.5.2.21) $\zeta = \eta \to \zeta \leqslant \eta$.

If ζ is identical with η, then ζ is part of η.

(2.5.2.22) $\zeta = \eta \to \zeta \circ \zeta$.

If ζ is identical with η, then ζ overlaps itself.

(2.5.2.23) $(At\zeta \mathbin{\&} \eta \leqslant \zeta) \to \eta = \zeta$.

A part of an atom is identical with that atom.

(2.5.2.24) $(\zeta \leqslant \eta \mathbin{\&} \sim \zeta < \eta) \to \zeta = \eta$.

If ζ is part, but not a proper part, of η then ζ is identical with η.

(2.5.2.25) $(At\zeta \mathbin{\&} \zeta \circ \eta) \to \zeta \leqslant \eta$.

An atom which overlaps an individual is part of the individual.

(2.5.2.26) $(At\zeta \mathbin{\&} At\eta \mathbin{\&} \zeta \circ \eta) \to \zeta = \eta$.

If two atoms overlap then they are identical.

(2.5.2.27) $\zeta < \eta \to (\exists \alpha)(At\alpha \mathbin{\&} \alpha \leqslant \eta \mathbin{\&} \sim \alpha \leqslant \zeta)$.

If ζ is a proper part of η then there is an atom which is part of η but not part of ζ.

(2.5.2.28) $(\exists \alpha) \alpha < \zeta \leftrightarrow (\exists \alpha)(\exists \beta)(At\alpha \mathbin{\&} At\beta \mathbin{\&} \sim \alpha = \beta$
$\mathbin{\&} \alpha < \zeta \mathbin{\&} \beta < \zeta)$.

An individual has a proper part if and only if it has at least two distinct atoms for proper parts.

(2.5.2.29) $(\zeta = \eta \mathbin{\&} \zeta \circ \tau) \to \eta \circ \tau$.

If ζ is identical with η and ζ overlaps τ, then η overlaps τ.

(2.5.2.30) $[\zeta \circ \zeta \ \& \ (\forall \alpha)([At\,\alpha \ \& \ \alpha \leqslant \zeta] \to \alpha \leqslant \eta)] \to \zeta \leqslant \eta$.

If every atom which is part of ζ is part of η, then ζ itself is part of η (provided that ζ exists).

(2.5.2.31) $\zeta = \eta \leftrightarrow (\zeta \circ \zeta \ \& \ \eta \circ \eta \ \& \ (\forall \alpha)[At\,\alpha \to (\alpha \leqslant \zeta \leftrightarrow \alpha \leqslant \eta)])$.

Individuals are identical if and only if they exist and have the same atomic parts. This principle of individuation is equivalent to the formula (2.4.5) which turned out not to be a theorem of LGCI.

(2.5.2.32) $[\zeta \circ \zeta \ \& \ (\forall \alpha)([At\,\alpha \ \& \ \alpha \circ \zeta] \to \alpha \circ \eta) \to \zeta \leqslant \eta$.

If ζ exists and if every atom which overlaps ζ overlaps η, then ζ is part of η.

(2.5.2.33) $\zeta = \eta \leftrightarrow (\zeta \circ \zeta \ \& \ \eta \circ \eta \ \& \ (\forall \alpha)[At\,\alpha \to (\alpha \circ \zeta \leftrightarrow \alpha \circ \eta)])$.

Individuals are identical if and only if they exist and overlap the same atoms.

(2.5.2.34) $(\forall \alpha)(\exists \beta)(At\,\beta \ \& \ \beta \leqslant \alpha)$.

Every (actual) individual has an atomic part.

(2.5.2.35) $At\,\zeta \leftrightarrow [\zeta \circ \zeta \ \& \ (\forall \alpha)(\zeta \circ \alpha \to \zeta \leqslant \alpha)]$.

Atoms are just those existing individuals which are part of everything which they overlap. (From left to right, use (2.5.2.25); from right to left, assume the hypothesis, assume that ζ is not an atom, and use D3, D2, 2.5.2.13, 2.5.2.9.)

(2.5.2.36) $At\,\zeta \leftrightarrow [\zeta \circ \zeta \ \& \ (\forall \alpha)(\forall \beta)([\alpha \circ \zeta \ \& \ \beta \circ \zeta] \to \alpha \circ \beta)]$.

ζ is an atom if and only if ζ exists and any two individuals which overlap ζ overlap one another.

Informally, from left to right, use (2.5.2.25) and (Ax3). From right to left, assume the hypothesis and suppose that ζ is not an atom. Hence, ζ has a proper part, say y. By (2.5.2.27), there is an atom, say z, which is part of ζ but not part of y. By (2.5.2.9), z overlaps ζ. Also, by (2.5.2.34), there exists an atom, say u, which is part of y, hence part of ζ, and therefore u overlaps ζ. Since both u and z overlap ζ, and by the hypothesis, they overlap one another; and since they are both atoms, by (2.5.2.26), $z = u$. But then, by (2.5.2.21) and (2.5.2.8), z is part of y: contradiction.

(2.5.2.37) $[\zeta \circ \zeta \ \& \ \sim At\,\zeta \ \& \ (\forall \alpha)(\alpha < \zeta \to \alpha \leqslant \eta)] \to \zeta \leqslant \eta$.

If ζ exists and is not an atom, and if every proper part of ζ is part of η, then ζ itself is part of η.

2.5.3. *The Semantics of CII*

We shall presently provide a model-theoretic interpretation of CII. In doing so, we refer to the definitions and informal remarks concerning semantics which occur in the second half of Section 1.3.2. We begin by stating and then briefly discussing the notion of a model:

(2.5.3.1) DEFINITIONS: **M** is a *model* of CII if and only if there exist **R**, **U**, and **A** such that $\mathbf{M} = \langle \mathbf{R}, \mathbf{U}, \mathbf{A} \rangle$ and
(1) **U** is an atomistic universe of individuals for **R**, and
(2) **A** is a function whose domain is a set of variables and whose range is included in **U**.
In this context, we call **R** *the part-whole relation*, **U** *the universe* (of discourse), and **A** *the assignment* (to variables) *of* the model **M**.

In appraising this definition, we recall the definitions of a part-whole relation (2.2.6) and of an atomistic universe of individuals for **R** (2.3.11), and note that being such a universe implies that **R** is a part-whole relation.

The third constituent, **A**, of a model assigns to some variables (not necessarily to all of them) individuals in **U**. Variables which are not in the domain of **A** are regarded as having *no values* in the model. We permit that the universe **U** of a model is empty; and in that event, the domain of **A** must be empty, and no variables whatever have values in the model. The plausibility of this treatment of variables will become clearer in subsequent chapters, where denotationless terms other than variables shall be admitted. Every model **M** is uniquely determined by its respective constituents **R**, **U**, and **A**.

We use the notation '$\mathbf{A}_x^\alpha$' to denote that assignment which differs from **A** itself, if at all, only by assigning x to α:

(2.5.3.2) DEFINITION: $\mathbf{A}_x^\alpha = [\mathbf{A} \sim \{(\alpha, \mathbf{A}(\alpha))\}] \cup \{(\alpha, x)\}$.

Thus, $\mathbf{A}_x^\alpha(\alpha) = x$; and for all items $\beta \neq \alpha$, $\mathbf{A}_x^\alpha(\beta) = \mathbf{A}(\beta)$.

For all formulas ϕ of CII containing only undefined symbols, we specify recursively under what conditions ϕ is satisfied by a model:

INDIVIDUALS 57

(2.5.3.3) DEFINITION: **M** *sat* ϕ [in words: **M** satisfies ϕ in CII] if and only if for some **R**, **U**, and **A**, **M** = $\langle$**R**, **U**, **A**$\rangle$ is a model of CII, and either:
(1) for some variables α and β, $\phi = \ulcorner \alpha \circ \beta \urcorner$ and there exists an x in **U** such that x bears **R** to both **A**(α) and **A**(β),
(2) for some formula ψ, $\phi = \ulcorner \sim \psi \urcorner$ and it is not the case that **M** sat ψ,
(3) for some formulas ψ, χ, $\phi = \ulcorner (\psi \to \chi) \urcorner$ and **M** sat ψ only if **M** sat χ, or
(4) for some formula ψ and variable α, $\phi = \ulcorner (\forall \alpha) \psi \urcorner$ and for all x in **U**, $\langle$**R**, **U**, **A**$_x^\alpha\rangle$ sat ψ.

By the conventions indicated in (1.3.1.38), **A**$(\alpha) = 0$ whenever α is not in the domain of **A**, and by the definitions, 0 is not in the field of **R** if **R** is a part-whole relation. Thus, we have:

(2.5.3.4) REMARK: If **M** sat $\ulcorner \alpha \circ \beta \urcorner$, then both α and β are in the domain of **A** and, of course, both **A**(α) and **A**(β) are in the universe of **M**.

Thus, the one primitive predicate '∘' of CII is interpreted by saying that formulas of the form $\ulcorner \alpha \circ \beta \urcorner$ are satisfied by a model just in case the values of α and β have a common **R**-part in the universe of that model. Since the universes in question are atomistic ones, this, in turn, amounts to saying that the values of α and β have a common **R**-least (or, atomic) part in the universe.

In accord with the remarks made in Section 1.3.2, we define:

(2.5.3.5) DEFINITION: ϕ is *valid* (in CII) just in case for all models **M**, **M** sat ϕ (in CII).

(2.5.3.6) DEFINITION: K (semantically) *yields* ϕ [or, the argument whose premises are the members of K and whose conclusion is ϕ is a valid one] just in case K is a class of formulas (of CII) and for every model **M**, if **M** satisfies (in CII) all members of K then **M** satisfies (in CII) ϕ.

In the next section, we shall sketch a proof of the following result:

(2.5.3.7) All valid formulas, and they alone, are theorems of CII.

Hence, roughly, the theorems of CII are true of every atomistic universe of individuals relative to any part-whole relation. And conversely, whenever a formula is true of every admissible universe, it can be derived as a theorem from the axioms of CII. More generally:

(2.5.3.8) If K is any set of formulas and ϕ is any formula, then ϕ is derivable from K just in case the argument whose premises are the members of K and whose conclusion is ϕ is a valid one.

Thus, the deductive machinery of CII is adequate to the given semantical interpretation. Once these results have been noted, the reader who is not interested in the details of the proofs may omit the next section.

2.5.4. *The Semantical Adequacy of CII**

A number of lemmas are stated whose proofs are either easy or well known, or have been treated elsewhere.[19] All references to CII are omitted.

(2.5.4.1) LEMMA: Suppose that $\mathbf{M} = \langle \mathbf{R}, \mathbf{U}, \mathbf{A} \rangle$ and $\mathbf{N} = \langle \mathbf{R}, \mathbf{U}, \mathbf{A}_x^\alpha \rangle$ are models and that α is a variable which is not free in ϕ. Then $\mathbf{M}$ sat ϕ if and only if $\mathbf{N}$ sat ϕ.

(2.5.4.2) LEMMA: Suppose that $\mathbf{M} = \langle \mathbf{R}, \mathbf{U}, \mathbf{A} \rangle$ and $\mathbf{N} = \langle \mathbf{R}, \mathbf{U}, \mathbf{A}_{\mathbf{A}(\beta)}^\alpha \rangle$ are models, α and β are variables, ϕ is a formula, and α is in the domain of $\mathbf{A}$. Then $\mathbf{M}$ sat ϕ_β^α if and only if $\mathbf{N}$ sat ϕ.

(2.5.4.3) LEMMA: If ϕ and ψ are formulas and α is a variable not free in ϕ, then $\ulcorner(\forall\alpha)[\phi \to \psi] \to [\phi \to (\forall\alpha)\psi]\urcorner$ is a theorem.

(2.5.4.4) LEMMA: If K and L are sets of formulas, $K \subseteq L$, and ϕ is derivable from K, then ϕ is derivable from L.

(2.5.4.5) LEMMA: If a formula ϕ is derivable from a set K of formulas, then ϕ is derivable from some finite subset of K.

(2.5.4.6) LEMMA (Deduction Theorem): If K is a set of formulas and ϕ, ψ are formulas, then ψ is derivable from $K \cup \{\phi\}$ if and only if $\ulcorner(\phi \to \psi)\urcorner$ is derivable from K.

(2.5.4.7) LEMMA: If ϕ is derivable from a set K of formulas and the variable α is not free in any member of K, then $\ulcorner(\forall\alpha)\phi\urcorner$ is derivable from K.

(2.5.4.8) LEMMA (Lindenbaum's Theorem): Every consistent set of

INDIVIDUALS 59

formulas is included in a maximally consistent set of formulas.

(2.5.4.9) LEMMA: If K is a consistent set of formulas and there are infinitely many variables which are not free in any member of K, then there exists a consistent set K' of formulas such that K is a subset of K' and for all formulas of the form $\ulcorner(\forall\alpha)\phi\urcorner$ there is a variable β such that $\ulcorner(\phi_\beta^\alpha \vee \sim\beta\circ\beta)\to(\forall\alpha)\phi\urcorner$ is in K'.

(2.5.4.10) THEOREM (Leibniz' Law): Suppose that ϕ' results from ϕ by replacing one or more free occurrences of α by β, and α is not free in any subformula of ϕ of the form $\ulcorner(\forall\beta)\psi\urcorner$ or $\ulcorner(\exists\beta)\psi\urcorner$. Then $\ulcorner(\alpha=\beta \& \phi)\to\phi'\urcorner$ is a theorem.

In stating subsequent theorems, we shall assume that the satisfaction conditions of (2.5.3.3) have been extended to take account of all logical constants and of all defined predicates of CII in conformity with the definitions (D1)–(D4).

(2.5.4.11) LEMMA: Suppose that $\mathbf{M} = \langle \mathbf{R}, \mathbf{U}, \mathbf{A} \rangle$ is a model. Then
(1) $\mathbf{M}$ sat $\ulcorner\alpha\leq\beta\urcorner$ just in case $\mathbf{A}(\alpha)\,\mathbf{R}\,\mathbf{A}(\beta)$,
(2) $\mathbf{M}$ sat $\ulcorner\alpha<\beta\urcorner$ just in case $\mathbf{A}(\alpha)\,\mathbf{R}\,\mathbf{A}(\beta)$ and not: $\mathbf{A}(\beta)\,\mathbf{R}\,\mathbf{A}(\alpha)$,
(3) $\mathbf{M}$ sat $\ulcorner\alpha=\beta\urcorner$ just in case $\mathbf{A}(\alpha)=\mathbf{A}(\beta)$, and
(4) $\mathbf{M}$ sat $\ulcorner At\alpha\urcorner$ just in case $\mathbf{A}(\alpha)$ is R-least in $\mathbf{U}$.

PROOF: (1) (a) Sufficiency: Assume that $\mathbf{M}$ sat $\ulcorner\alpha\leq\beta\urcorner$. By (D1), $\mathbf{M}$ sat $\ulcorner\alpha\circ\beta\urcorner$, by (2.5.3.4), $\mathbf{A}(\alpha)$, $\mathbf{A}(\beta)$ are in $\mathbf{U}$, and $\mathbf{A}(\beta)$ is in the field of $\mathbf{R}$. Show (aa): every R-least element of $\mathbf{U}$ which bears $\mathbf{R}$ to $\mathbf{A}(\alpha)$ bears $\mathbf{R}$ to $\mathbf{A}(\beta)$. Assume that z is R-least in $\mathbf{U}$ and $z\,\mathbf{R}\,\mathbf{A}(\alpha)$. Then $\langle \mathbf{R}, \mathbf{U}, \mathbf{A}_z^\gamma \rangle$ sat $\ulcorner\gamma\circ\alpha\urcorner$. By reasoning conforming to (D1), $\langle \mathbf{R}, \mathbf{U}, \mathbf{A}_z^\gamma \rangle$ sat $\ulcorner\gamma\circ\beta\urcorner$. Hence, for some x in $\mathbf{U}$, $x\mathbf{R}z$ and $x\,\mathbf{R}\,\mathbf{A}(\beta)$. Since z is R-least in $\mathbf{U}$, $x=z$. Hence, $z\,\mathbf{R}\,\mathbf{A}(\beta)$. By lemma (2.3.12), $\mathbf{A}(\alpha)\,\mathbf{R}\,\mathbf{A}(\beta)$.

(b) Necessity: Assume that $\mathbf{A}(\alpha)\,\mathbf{R}\,\mathbf{A}(\beta)$. Since 0 is not in the field of $\mathbf{R}$, $\mathbf{A}(\alpha)$ is in $\mathbf{U}$ and, clearly, $\mathbf{M}$ sat $\ulcorner\alpha\circ\alpha\urcorner$. We show that $\mathbf{M}$ sat $\ulcorner(\forall\gamma)(\gamma\circ\alpha\to\gamma\circ\beta)\urcorner$. Assume that x is in $\mathbf{U}$ and that $\langle \mathbf{R}, \mathbf{U}, \mathbf{A}_x^\gamma \rangle$ sat $\ulcorner\gamma\circ\alpha\urcorner$. Hence, for some y in $\mathbf{U}$, $y\mathbf{R}x$ and $y\mathbf{R}\,\mathbf{A}(\alpha)$. By the transitivity of $\mathbf{R}$, $y\,\mathbf{R}\,\mathbf{A}(\beta)$. Thus, $\langle \mathbf{R}, \mathbf{U}, \mathbf{A}_x^\gamma \rangle$ sat $\ulcorner\gamma\circ\beta\urcorner$. $\mathbf{M}$ satisfies the definiens of (D1), and hence $\mathbf{M}$ sat $\ulcorner\alpha\leq\beta\urcorner$. The parts (2), (3), and (4) of the lemma are obvious, given (1) and the definitions.

(2.5.4.12) LEMMA: For all variables α, β, and γ such that $\gamma \neq \alpha$ and $\gamma \neq \beta$, $\ulcorner \alpha \circ \beta \rightarrow (\exists \gamma)(At\,\gamma\, \&\, \gamma \leqslant \alpha\, \&\, \gamma \leqslant \beta)\urcorner$ is valid.

PROOF: Assume that $\mathbf{M} = \langle \mathbf{R}, \mathbf{U}, \mathbf{A} \rangle$ is a model and that $\mathbf{M}$ sat $\ulcorner \alpha \circ \beta \urcorner$. Then, for some x in $\mathbf{U}$, $x\,\mathbf{R}\,\mathbf{A}(\alpha)$ and $x\,\mathbf{R}\,\mathbf{A}(\beta)$. By the definition (2.3.11) of a universe, there exists a set S such that all members of S are $\mathbf{R}$-least in $\mathbf{U}$ and $x = \sup_\mathbf{R} S$. Note that $S \neq 0$ (for otherwise x, being in the field of $\mathbf{R}$, would equal 0; which is impossible). Let y be in S. Since y is $\mathbf{R}$-least in $\mathbf{U}$ and by Lemma (2.5.4.11), $\langle \mathbf{R}, \mathbf{U}, \mathbf{A}_y^\gamma \rangle$ sat $\ulcorner At\,\gamma \urcorner$. Also, $y\mathbf{R}x$ and hence y bears $\mathbf{R}$ to both $\mathbf{A}(\alpha)$ and $\mathbf{A}(\beta)$. Thus, $\langle \mathbf{R}, \mathbf{U}, \mathbf{A}_y^\gamma \rangle$ sat $\ulcorner(\gamma \leqslant \alpha\,\&\,\gamma \leqslant \beta)\urcorner$ by (2.5.4.11). This completes the proof.

(2.5.4.13) THEOREM (Soundness): All of the axioms (L1)–(L3) and (Ax1)–(Ax3) of CII are valid, and the inference rules of CII preserve validity.

The proof of this result is straightforward. One conditional of (Ax3) is valid due to the preceding lemma. In order to verify the converse conditional, it suffices to verify that the symmetry of '$\circ$' is valid. Even though we permit empty universes, there is no difficulty in showing that Modus Ponens preserves validity.

(2.5.4.14) LEMMA: Suppose that K is a consistent set of formulas and there are infinitely many variables which are not free in any member of K. Then there exists a model $\mathbf{M}$ such that $\mathbf{M}$ satisfies every member of K.

PROOF: Assume the hypothesis of the lemma.

(1) By Lemma (2.5.4.9), there exists a consistent extension K' of K such that for all formulas of the form $\ulcorner(\forall \alpha)\phi\urcorner$ there is a variable β such that $\ulcorner(\phi_\beta^\alpha \vee \sim \beta \circ \beta) \rightarrow (\forall \alpha)\phi\urcorner$ is in K'.

(2) By Lemma (2.5.4.8), there exists a maximally consistent extension K^* of K'.

(3) Let $A_0 = $ the set of all unit-sets of variables.

(4) Let $F = $ the set of all unions of non-empty subsets of A_0.

(5) Let $\mathbf{R} = $ inclusion restricted to F.

It is easily verified that $\mathbf{R}$ satisfies the conditions imposed by (2.2.6) on part-whole relations. Note that for any S in F, $\sup_\mathbf{R} S = $ the union of S.

(6) Let $\mathbf{U} = $ the set of all items x such that $x \neq 0$ and for some variable α, $x = $ the set of all variables β such that $\ulcorner At\,\beta\,\&\,\beta \leqslant \alpha \urcorner$ is in K^*.

(7) Let $\mathbf{A} = $ that function whose domain is the set of all variables α

such that ⌜α∘α⌝ is in K^* and for every such α, $A(α)$=the set of all variables β such that ⌜At β & β ⩽ α⌝ is in K^*.

(8) We show that **U** is an atomistic universe of individuals for **R**.

(a) Clearly, **U** is a subset of the field F of **R**.

(b) Show: for every x in **U** there is a set S of **R**-least elements in **U** such that $x=$ the union of S. Assume $x \in$ **U**. Hence $x \neq 0$ and for some α, $x=\{β: ⌜At β & β ⩽ α⌝ \in K^*\}$.

Let $S =$ the set of all items y such that for some $β \in x$, $y=\{γ: ⌜γ = β⌝ \in K^*\}$. In order to show that every member of S is **R**-least in **U**, assume that $y \in S$ and, to the contrary, that y is not **R**-least in U. Hence, for some $z \in U$, $z \subset y$. For some variable ζ, $z = \{δ: ⌜At δ & δ ⩽ ζ⌝ \in K^*\}$. Since $z \subset y$, for some γ, $γ \in y$ and $γ \notin z$. For some $β \in x$, ⌜γ = β⌝ $\in K^*$ (since $γ \in y$). ⌜At β⌝ $\in K^*$ and, by Leibniz' Law (2.5.4.10), ⌜At γ⌝ $\in K^*$.

On the other hand, since $z \in$ U, $z \neq 0$ and for some δ, ⌜At δ & δ ⩽ ζ⌝ $\in K^*$. $δ \in z$; hence $δ \in y$; thus ⌜δ = β⌝ $\in K^*$. By transitivity of '=', ⌜γ = δ⌝ $\in K^*$. Therefore, by (2.5.4.10), ⌜At γ & γ ⩽ ζ⌝ $\in K^*$. Thus, $γ \in z$, contrary to the earlier assertion.

Also, $y = \{γ: ⌜At γ & γ ⩽ β⌝ \in K^*\}$. Hence, $y \in U$. Since it is a theorem that ⌜At γ → γ = γ⌝, and by (2.5.4.10), clearly, $x = \bigcup S$.

(9) Let **M** = ⟨**R, U, A**⟩. **M** is a model.

(10) Show: for all formulas ϕ, **M** sat ϕ if and only if $ϕ \in K^*$.

We use induction on the length of primitive formulas.

(a) $ϕ = $ ⌜α ∘ β⌝, for some variables α and β.

Sufficiency: Assume that **M** sat ⌜α ∘ β⌝. Hence for some x in **U**, $x \subseteq A(α)$ and $x \subseteq A(β)$. $x \neq 0$ and for some δ, $x = \{γ: ⌜At γ & γ ⩽ δ⌝ \in K^*\}$. Let $γ \in x$, so that $γ \in A(α)$ and $γ \in A(β)$. Recalling how '**A**' was defined in (7), ⌜At γ & γ ⩽ α⌝ and ⌜At γ & γ ⩽ β⌝ $\in K^*$. By (Ax3), ⌜α ∘ β⌝ $\in K^*$.

Necessity: Assume that ⌜α ∘ β⌝ $\in K^*$. By (Ax3) and (1), for some γ, ⌜At γ & γ ⩽ α & γ ⩽ β⌝ $\in K^*$. Both, ⌜α ∘ α⌝ and ⌜β ∘ β⌝ are in K^*, so that both α and β are in the domain of **A**.

But $A(γ) \subseteq A(α)$; for, suppose $δ \in A(γ)$, then by (7) ⌜At δ & δ ⩽ γ⌝ $\in K^*$ and, by transitivity of '⩽', ⌜δ ⩽ α⌝ $\in K^*$ which implies that $δ \in A(α)$. Similarly, $A(γ) \subseteq A(β)$. Since ⌜γ ∘ γ⌝ $\in K^*$ and by (7), $A(γ) \in U$. Hence **M** sat ⌜α ∘ β⌝.

(b) The sentential steps of this inductive argument present no difficulties, and the universal case, appealing repeatedly to (1), proceeds almost as usual.[20]

(11) For all formulas $\phi \in K$, **M** sat ϕ.

(2.5.4.15) THEOREM: If K is a consistent set of formulas, then there exists a model **M** such that **M** satisfies every member of K.

In proving this theorem, on the basis of Lemma (2.5.4.14), one proceeds in the manner customary to treatments of the predicate calculus.

(2.5.4.16) THEOREM (Strong Completeness): If K is a set of formulas and ϕ is a formula, then K (semantically) yields ϕ only if ϕ is derivable from K.

This theorem is an immediate consequence of the preceding one. If, in the preceding theorem, we set $K=0$, we obtain:

(2.5.4.17) THEOREM (Limited Completeness): Every valid formula is a theorem.

Although the particular part-whole relation employed in the proof of (2.5.4.14) happened to be that of inclusion, it is clear (but will not be proved here) that for any part-whole relation **R**, every consistent class of formulas has a model whose part-whole relation is just **R**.

Thus, the semantical adequacy of CII is assured.

2.6. FINITE SUMS AND PRODUCTS: THE CALCULUS CIII

As we have briefly indicated in Section 2.4, CIII is intended to be a calculus treating of atomistic universes of individuals which are such that the intersection of overlapping individuals and the sum of any two individuals is always again an individual. Although it has seemed to us, in Section 2.3, that there are part-whole relations with respect to which sums of individuals should not always in turn be individuals, the special universes of which that is true can serve to interpret a somewhat stronger theory of individuals.

An auxiliary notion is required: that of a product of two objects relative to an arbitrary partial ordering. The notion is defined in analogy to that of a greatest lower bound of two numbers relative to 'less than or equal to', or to that of a greatest lower bound (intersection, or product) of two sets relative to inclusion. The more general notion of a product or infimum, the counterpart of the 'infinite' operation of a supremum given in Definition (2.2.1), is not required in our development.

INDIVIDUALS

(2.6.1) DEFINITION: $\inf_R(x, y)$ [the *infimum*, relative to R, of x and y] = the unique object z which is such that (1) z bears R to both x and y, (2) for all u, if u bears R to both x and y, then u bears R to z, and (3) R is a partial ordering. We say that $\inf_R(x, y)$ *exists* just in case there is one and only one object z such that the conditions (1), (2), and (3) are satisfied.

The infimum of two objects (if it exists) is just the supremum (if it exists) of their common parts; that is to say:

(2.6.2) REMARK: Let S = the set of all items z, such that zRx and zRy, and suppose that $\inf_R(x, y)$ and $\sup_R S$ both exist. Then $\inf_R(x, y) = \sup_R S$.

In presenting the formal system CIII, we refer to the appropriate section in the formalization of CII:

2.6.1. *The Vocabulary and the Axioms of CIII*

The *vocabulary* of CIII, and its division into primitive and defined constants, is that of CII.

All axioms of CII (both definitional and proper) are also *axioms* of CIII. In addition, for all terms [21] ζ and η and for all distinct variables α and β not occurring in either ζ or η, the following are axioms of CIII:

(Ax4) $\ulcorner \zeta \circ \eta \to (\exists \beta)(\forall \alpha)(At\,\alpha \to [\alpha \leqslant \beta \leftrightarrow (\alpha \leqslant \zeta\, \&\, \alpha \leqslant \eta)])\urcorner$.

In words: If two individuals ζ and η overlap, then there exists an individual β (the product of ζ and η) such that all and only those atoms are part of β which are common parts of ζ and η. Informally: the product of any two overlapping individuals exists.

(Ax5) $\ulcorner (\zeta \circ \zeta\, \&\, \eta \circ \eta)$
$\to (\exists \beta)(\forall \alpha)(At\,\alpha \to [\alpha \leqslant \beta \leftrightarrow (\alpha \leqslant \zeta \vee \alpha \leqslant \eta)])\urcorner$.

In words: If the individuals ζ and η overlap themselves (or, exist), then there is an individual β (the sum of ζ and η) such that all and only those atoms are part of β which are parts of either ζ or η. Informally: the sum of any two actual individuals exists.

The *theorems* of CIII comprise those of CII, together with those formulas which are derivable by application of the inference rules to (Ax4) and (Ax5). At present we are not interested in these additional theorems.

2.6.2. *The Semantics of CIII*

We begin by defining and then discussing the notion of a model:

(2.6.2.1) DEFINITION: M is a *model* of CIII if and only if there exist R, U, and A such that $M = \langle R, U, A \rangle$ is a model of CII, and, in addition, the following two conditions are satisfied:
(1) for all x and y in U: if something bears R to both x and y, then $\inf_R(x, y)$ is a member of U, and
(2) for all x and y in U: $\sup_R\{x, y\}$ is a member of U.[22]

The condition (1) of this definition can be paraphrased by saying: If two individuals (actualized in U) have a common R-part (whether or not that part is actualized in U), then their infimum or product (relative to R) is also actualized in U. This principle has the consequence, more precisely expressed below, that distinct atoms in U are discrete:

(2.6.2.2) REMARK: Suppose that $\langle R, U, A \rangle$ is a model of CIII, and that x and y are distinct R-least elements of U. Then there is no item z such that zRx and zRy.

In order to illustrate the additional strength bestowed by condition (1) on universes of individuals, we revert to the part-whole relation described in Example (2.3.4); namely, the inclusion relation between non-empty unions of singletons. Consider the following universes of individuals relative to inclusion:

$$U_1 = \{\{A, B\}, \{B, C\}\},$$
$$U_2 = \{\{A, B\}, \{B, C\}, \{C, D\}\}.$$

All constituents in each universe are atomic in that universe. But the atoms in U_1 are not 'discrete' in the sense that $\{B\}$ is included in both of them. Here the common part, $\{B\}$, is not itself in the universe. But the atoms $\{A, B\}$ and $\{C, D\}$ of U_2 fail to be 'discrete' even in the sense that an atom of the universe itself, $\{B, C\}$ is included in the sum of the other two. Both U_1 and U_2, although they are atomistic universes of individuals, are not universes which satisfy the condition (1) of the definition.

The condition (2) of the last definition (that the sum or supremum, relative to **R**, of any two individuals which are actualized in **U** is always also actualized in **U**) is self-explanatory. The universes U_1 and U_2 above also exemplify atomistic universes of individuals which fail to meet condition (2).

Apart from our new notion of a model, all semantical concepts of CIII are exactly similar to the ones of CII.

To readers who are interested in proofs, we shall demonstrate in the next section that all and only those formulas which are derivable from the axioms of CIII are satisfied in every model of CIII.

2.6.3. *The Semantical Adequacy of CIII**

All of the lemmas and theorems stated in Section 2.5.4 regarding the calculus CII continue to hold of CIII. We omit, in this section, all references to CIII.

(2.6.3.1) LEMMA: For all distinct variables α, β, γ, and δ, the formula $\ulcorner \alpha \circ \beta \to (\exists \gamma)(\forall \delta)(At\delta \to [\delta \leqslant \gamma \leftrightarrow (\delta \leqslant \alpha \,\&\, \delta \leqslant \beta)])\urcorner$ is valid; that is, (Ax4) is valid.

PROOF: Assume that $\mathbf{M} = \langle \mathbf{R}, \mathbf{U}, \mathbf{A} \rangle$ is a model and that **M** sat $\ulcorner \alpha \circ \beta \urcorner$. Then, for some x in **U**, x **R** $\mathbf{A}(\alpha)$ and x **R** $\mathbf{A}(\beta)$. Both $\mathbf{A}(\alpha)$ and $\mathbf{A}(\beta)$ are in **U** (for otherwise they would be empty and could not be in the field of **R**). By condition (1) on models, $\inf_\mathbf{R}(\mathbf{A}(\alpha), \mathbf{A}(\beta))$ is in **U**. Let $y = \inf_\mathbf{R}(\mathbf{A}(\alpha), \mathbf{A}(\beta))$. Assigning y to γ, we show that the consequent of (Ax4) is satisfied. Let $\mathbf{B} = \mathbf{A}_{yz}^{\gamma\delta}$ and $\mathbf{N} = \langle \mathbf{R}, \mathbf{U}, \mathbf{B} \rangle$; and assume that $z \in \mathbf{U}$, and that **N** sat $\ulcorner At\, \delta \urcorner$. By Lemma (2.5.4.11), z is **R**-least in **U**.

Sufficiency: Show, if **N** sat $\ulcorner \delta \leqslant \gamma \urcorner$ then **N** sat $\ulcorner \delta \leqslant \alpha \,\&\, \delta \leqslant \beta \urcorner$. Assume the hypothesis. By Lemma (2.5.4.11), part (1) of the definition (2.6.1) and transitivity, the consequent follows immediately.

Necessity: Show, if **N** sat $\ulcorner \delta \leqslant \alpha \,\&\, \delta \leqslant \beta \urcorner$ then **N** sat $\ulcorner \delta \leqslant \gamma \urcorner$. Assume the hypothesis. By (2.5.4.11), $\mathbf{B}(\delta)$ **R** $\mathbf{B}(\alpha)$ and $\mathbf{B}(\delta)$ **R** $\mathbf{B}(\beta)$. Hence, by part (2) of the definition (2.6.1), $\mathbf{B}(\delta)$ **R** $\mathbf{B}(\gamma)$.

(2.6.3.2) LEMMA: Suppose that **R** is a part-whole relation, **A** is the set of **R**-least elements, S is a non-empty subset of the field

of R, and $T=\{a: a\in A$ and for some y in S, $aRy\}$. Then
(1) $\sup_R S = \sup_R T$, and
(2) if $x = \sup_R S$, $a\in A$, and aRx, then for some y in S, aRy.

(2.6.3.3) LEMMA: For all distinct variables α, β, γ, and δ, the formula
⌜$(\alpha \circ \alpha \& \beta \circ \beta) \to (\exists \gamma)(\forall \delta)(At\delta \to [\delta \leqslant \gamma \leftrightarrow (\delta \leqslant \alpha \vee \delta \leqslant \beta)])$⌝
– that is, (Ax5) – is valid.

PROOF: Assume that $\mathbf{M} = \langle \mathbf{R}, \mathbf{U}, \mathbf{A}\rangle$ is a model and that M sat ⌜$\alpha \circ \alpha \& \beta \circ \beta$⌝. Hence, by the satisfaction condition regarding '∘', both $\mathbf{A}(\alpha)$ and $\mathbf{A}(\beta)$ are in the field of R, non-empty, and hence in U. Let $z = \sup_R\{\mathbf{A}(\alpha), \mathbf{A}(\beta)\}$. By (2) of Definition (2.6.2.1), $z\in U$, and hence the supremum exists. In order to show that M satisfies the consequent of (Ax5), let $\mathbf{B} = A_{zx}^{\gamma\delta}$, let $\mathbf{N} = \langle \mathbf{R}, \mathbf{U}, \mathbf{B}\rangle$, and assume $x\in U$ and that N sat ⌜$At\, \delta$⌝, so that x is R-least in U (by 2.5.4.11).

Sufficiency: Show, if N sat ⌜$\delta \leqslant \gamma$⌝ then N sat ⌜$\delta \leqslant \alpha \vee \delta \leqslant \beta$⌝. It suffices to show that if xRz then either $xR\mathbf{A}(\alpha)(=\mathbf{B}(\alpha))$ or $xR\mathbf{A}(\beta)(=\mathbf{B}(\beta))$. Assume that xRz. Let $A=$ the set of R-least elements, and let $T = \{a: a\in A$ and either $aR\, \mathbf{A}(\alpha)$ or $a\, \mathbf{R}\, \mathbf{A}(\beta)\}$. By Lemma (2.6.3.2) (1), letting S there be $\{\mathbf{A}(\alpha), \mathbf{A}(\beta)\}$, $z = \sup_R T$. Since $x\in U$, for some $a\in A$, aRx. By transitivity, aRz. By Definition (2.2.6) (3a), $a\in T$. Hence, $a\, \mathbf{R}\, \mathbf{A}(\alpha)$ or $a\, \mathbf{R}\, \mathbf{A}(\beta)$. Since $\mathbf{A}(\alpha), \mathbf{A}(\beta)$ are in U and thus suprema of sets of atoms, and by Lemma (2.6.3.2) (2), there exists an R-least element v in U such that aRv and either $v\, \mathbf{R}\, \mathbf{A}(\alpha)$ or $v\, \mathbf{R}\, \mathbf{A}(\beta)$.

To follow up one of two similar cases, assume that $v\, \mathbf{R}\, \mathbf{A}(\alpha)$. Since both aRx and aRv, and by the condition (1) on models, $\inf_R(x, v)$ is in U, $\inf_R(x, v)$ bears R to v. In fact, since v is R-least in U, $\inf_R(x, v) = v$. Similarly, $\inf_R(x, v) = x$. Hence, $x = v$ and, since $vR\, \mathbf{A}(\alpha)$, $x\, \mathbf{R}\, \mathbf{A}(\alpha)$.

Necessity: Show, if N sat ⌜$\delta \leqslant \alpha \vee \delta \leqslant \beta$⌝ then N sat ⌜$\delta \leqslant \gamma$⌝. This is obvious by Lemma (2.5.4.11), the definition of suprema, and transitivity.

In proving the soundness of (Ax5), we had to know both: that universes are closed under suprema and that they are closed under infima. I do not know whether a theory whose models are closed only under suprema, but not under infima (in the given sense) can be axiomatized.

The essential step in demonstrating the completeness of CII was the proof of Lemma (2.5.4.14). In order to demonstrate the completeness of CIII, it is sufficient to prove that the model constructed in that earlier proof also satisfies the closure conditions imposed on models of CIII,

INDIVIDUALS 67

once the new axioms (Ax4) and (Ax5) are added. In order to save space, we refer by number back to the respective steps of that earlier proof.

In the proof of Lemma (2.5.4.14), let the steps (1)–(7) and (8a), (8b) stand as before. We insert the following steps:

(8c) Show: for all x and y in U, if something bears R to both x and y, then $\inf_R(x, y) \in U$. Assume that x and y are in U and that zRx and zRy. Recalling that R is inclusion and that R-least elements are singletons of variables, we observe that for some $u = \{\gamma\}, u \subseteq z, z \subseteq x$ and $z \subseteq y$. Let $x = \{\gamma: \ulcorner At\, \gamma\, \&\, \gamma \leqslant \alpha \urcorner \in K^*\}$ and $y = \{\gamma: \ulcorner At\, \gamma\, \&\, \gamma \leqslant \beta \urcorner \in K^*\}$. Thus, $\ulcorner At\, \gamma\, \&\, (\gamma \leqslant \alpha\, \&\, \gamma \leqslant \beta) \urcorner \in K^*$; hence, by (Ax3), $\ulcorner \alpha \circ \beta \urcorner \in K^*$. By (1) and (Ax4), for some variable δ, $\ulcorner (\forall \zeta)(At\, \zeta \to [\zeta \leqslant \delta \leftrightarrow (\zeta \leqslant \alpha\, \&\, \zeta \leqslant \beta)]) \urcorner$ is in K^*. Let $v = \{\gamma: \ulcorner At\, \gamma\, \&\, \gamma \leqslant \delta \urcorner \in K^*\}$. It is not hard to show that $v = x \cap y$. But, given what R is, $\inf_R(x, y) = x \cap y$. One may verify that $v \neq 0$ since both x and y are in U. Thus, $\inf_R(x, y) \in U$.

(8d) Show: for all x and y in U, $\sup_R\{x, y\} \in U$.

Assume that $x, y \in U$ and let x and y be as in (8c). Since $x, y \neq 0$, $\ulcorner \alpha \circ \alpha \urcorner, \ulcorner \beta \circ \beta \urcorner \in K^*$. By (Ax5) and (1), for some variable δ, $\ulcorner (\forall \zeta)(At\, \zeta \to [\zeta \leqslant \delta \leftrightarrow (\zeta \leqslant \alpha \vee \zeta \leqslant \beta)]) \urcorner \in K^*$. Let $z = \{\gamma: \ulcorner At\, \gamma\, \&\, \gamma \leqslant \delta \urcorner \in K^*\}$ $z \neq 0$ and $z \in U$. It follows easily that z is just the union of x and y which, in turn, is the desired supremum.

By continuing the proofs as indicated in connection with CII, we establish the semantical completeness of CIII.

2.7. INFINITE SUMS AND PRODUCTS: THE CALCULUS CI III

The interpreted formal system CIIII is to be such that its theorems hold in all those models of CIII (as defined by 2.6.2.1) which satisfy the additional requirement that their universes are closed, roughly speaking, under sums of their non-empty specifiable subsets.

In order to state this condition, we need the auxiliary notion of a specifiable (or definable) subset of a universe of individuals. Informally, this is to be a subset of a universe which can be specified, and its members determined, by some formula of the object-language of the system. Since the vocabulary of the formal systems under consideration (including the set of variables) is denumerable, the set of distinct formulas of these systems is also denumerable. Hence, it can be expected that at most denumerably many subsets of any given universe of individuals are specifiable by formulas of these systems. But the number of all subsets of a given universe of individuals may, of course, be greater than that.

A realist may hold that sets of individuals (in a given universe) are universals; and a nominalist, recalling the medieval doctrine that universals are mere *nomina* ('nominalism', from *'nominalis'*: 'pertaining to names'), might wish to identify these entities with certain of their names. The 'names' of sets of individuals, in turn, might be identified with those formulas which serve to define the sets in question. From this point of view, and given our object-languages, a nominalist can succeed in identifying with their 'names' at most denumerably many of those entities which a realist would be willing to endorse. A nominalist who is prepared to accept sums of individuals as new individuals, may well restrict his acceptance to sums of those sets of individuals which he succeeds in identifying with their 'names'; that is to say, he will endorse as new individuals only the sums of specifiable sets of individuals. These considerations suggest that our new condition on universes of individuals may be of some philosophical interest.

The precise notion of a specifiable set of individuals is defined with reference to the models of CII as follows:

(2.7.1) DEFINITION: S is *specifiable* in $\mathbf{M}$ (relative to CII) if and only if there exist $\mathbf{R}$, $\mathbf{U}$, $\mathbf{A}$, a variable α, and a formula ϕ such that $\mathbf{M} = \langle \mathbf{R}, \mathbf{U}, \mathbf{A} \rangle$, $\mathbf{M}$ is a model of CII, and for all x: $\langle \mathbf{R}, \mathbf{U}, \mathbf{A}_x^\alpha \rangle$ sat ϕ just in case $x \in S$.

Informally, and under omission of relativizations, a set S of individuals is specifiable if the language contains some formula ϕ (typically with one free variable) such that all and only those individuals are in S which satisfy that formula. For example, the entire universe of individuals is specifiable by formulas of the form '$x \circ x$', since all actual individuals satisfy that formula. And the set of all $\mathbf{R}$-least elements in the universe is specifiable by formulas of the form 'Atx', since that formula is satisfied by (or true of) just those individuals which are atomic in the universe.

In formalizing CIIII, we refer to the presentations of earlier systems:

2.7.1. *The Vocabulary and the Axioms of CIIII*

The *vocabulary* of CIIII, and the distinction between primitive and defined constants, is that of CII (and hence also that of CIII).

All axioms of CII (logical, definitional, and proper ones) are also *axioms* of CIIII. However, the axioms (Ax4) and (Ax5) peculiar to CIII, rather than having the status of axioms in CIIII, will turn out to be theorems of CIIII. In addition, for all distinct variables α and β, and for all formulas ϕ such that β does not occur in ϕ, the following are proper axioms of CIIII:

(AxS6) $\ulcorner(\exists\alpha)(At\,\alpha\ \&\ \phi) \to (\exists\beta)(\forall\alpha)(At\,\alpha \to [\alpha \leq \beta \leftrightarrow \phi])\urcorner.$

In words: If there exists an atom satisfying the (specifiable) condition ϕ, then there exists an individual β having all and only those atoms for parts which satisfy the condition ϕ. Informally: If some atom satisfies the condition ϕ, then the sum of all atoms satisfying ϕ exists.

2.7.2. *Some Theorems of CIIII*

All theorems of CII are theorems of CIIII. In addition, the axioms (Ax4) and (Ax5) peculiar to CIII are theorems of CIIII.

To derive (Ax4), consider (AxS6) and let $\phi = \ulcorner \alpha \leq \zeta\ \&\ \alpha \leq \eta \urcorner$. In proving (Ax5) within CIIII, let $\phi = \ulcorner \alpha \leq \zeta\ \vee\ \alpha \leq \eta \urcorner$ in (AxS6).

Thus, every theorem of CIII is also a theorem of CIIII.

Let us compare the reconstructed Leonard-Goodman system LGCI with CIIII. All logical axioms (L1)–(L3) are logical axioms of CIIII. The axiom (L4) of LGCI is a theorem of CII (namely, 2.5.4.10), and hence also a theorem of CIIII. The proper axioms (A1) and (A2) of LGCI are the respective axioms (Ax1) and (Ax2) of CII and thus of CIIII. The axiom (A3) of LGCI is Theorem (2.5.2.7) of CII. (A4) is a consequence of the definition (D4). (A5) is the same as Theorem (2.5.2.20). And the axiom-scheme of summation (AS6) of LGCI is derivable from the corresponding axiom-scheme (AxS6) of CIIII (by replacing 'ϕ' in (AxS6) by $\ulcorner(\exists\gamma)(\alpha \leq \gamma\ \&\ \phi_\gamma^\alpha)\urcorner$). Thus, all theorems of LGCI are theorems of CIIII. In addition, CIIII, but not LGCI, has for a theorem the syntactical counterpart (2.5.2.31) of the principle of individuation which was taken to be appropriate to individuals. Thus, CIIII has somewhat greater deductive power than LGCI.

2.7.3. *The Semantics of CIIII*

The theorems of CIIII are taken to be necessarily true of individuals which satisfy the conditions that the product of any two overlapping individuals and the sum of any specifiable non-empty set of individuals is always again an individual. Accordingly, we define the notion of a model for CIIII as follows:

(2.7.3.1) DEFINITION: **M** is a *model* of CIIII if and only if there exist **R**, **U**, and **A** such that $\mathbf{M} = \langle \mathbf{R}, \mathbf{U}, \mathbf{A} \rangle$ is a model of CII, and, in addition, the following two conditions are satisfied:
(1) for all x and y in **U**: if something bears **R** to both x and y, then $\inf_\mathbf{R}(x, y)$ is a member of **U**, and
(2) for all S: if S is non-empty and specifiable in **M** (relative to CII), then $\sup_\mathbf{R} S$ is a member of **U**.

The intuitive content of CIIII-models, additional to that of CIII-models discussed in the preceding section, is the following: Consider any class of individuals (within the universe) which is specifiable; that is, roughly speaking, which can be described by some condition expressible in the object-language of the present system. Such a class of individuals need not be finite. The sum of all individuals in any such (non-empty) class, finite or infinite, is again an individual in the universe of a CIIII-model. But in the universes of CII-models, only sums of finitely many individuals are in turn individuals. In particular, the sum of all individuals in the universe (which might be called 'the universal individual') will always exist in universes of CIIII, but need not exist in the universes of CIII.

All other semantical concepts of CIIII are exactly like those of CII.

The theory CIIII, like its two predecessors, is semantically both sound and complete, as we shall show in the next section.

2.7.4. *The Semantical Adequacy of CIIII**

In order to demonstrate the soundness of CIIII, it will suffice to prove the validity of (AxS6). Repeated use will be made of Lemma (2.5.4.11), which continues to hold in CIIII; and appeals to this lemma will not be made explicit.

(2.7.4.1) LEMMA: For all distinct variables α and β, and for all formulas ϕ such that β does not occur in ϕ, the formula $\ulcorner(\exists\alpha)(At\alpha \& \phi) \to (\exists\beta)(\forall\alpha)(At\alpha \to [\alpha \leq \beta \leftrightarrow \phi])\urcorner$, – that is, (AxS6) – is valid.

PROOF: Let $\mathbf{M} = \langle \mathbf{R}, \mathbf{U}, \mathbf{A} \rangle$ be a model, and suppose that $\mathbf{M}$ satisfies the hypothesis of (AxS6). Thus, for some $x \in \mathbf{U}$, $\langle \mathbf{R}, \mathbf{U}, \mathbf{A}_x^\alpha \rangle$ sat $\ulcorner At\,\alpha \& \phi \urcorner$, so that x is $\mathbf{R}$-least in $\mathbf{U}$.

Let $S = \{z : \langle \mathbf{R}, \mathbf{U}, \mathbf{A}_z^\alpha \rangle \text{ sat } \ulcorner At\,\alpha \& \phi \urcorner\}$.

$S \neq 0$. Since the formulas and models of CIIII are also formulas and models of CII, $\langle \mathbf{R}, \mathbf{U}, \mathbf{A}_z^\alpha \rangle$ sat $\ulcorner At\,\alpha \& \phi \urcorner$ in the sense of CIIII if and only if $\langle \mathbf{R}, \mathbf{U}, \mathbf{A}_z^\alpha \rangle$ sat $\ulcorner At\,\alpha \& \phi \urcorner$ in the sense of CII. Hence, S is specifiable in $\mathbf{M}$ and, by the condition (2) on models, $\sup_\mathbf{R} S \in \mathbf{U}$.

In order to show that M satisfies the consequent of (AxS6), let $y = \sup_\mathbf{R} S$, let $\mathbf{B} = \mathbf{A}_{yu}^{\beta\alpha}$, $\mathbf{N} = \langle \mathbf{R}, \mathbf{U}, \mathbf{B} \rangle$, assume that $u \in \mathbf{U}$ and that $\mathbf{N}$ sat $\ulcorner At\,\alpha \urcorner$. Hence, u is $\mathbf{R}$-least in $\mathbf{U}$.

Sufficiency: Show, if $\mathbf{N}$ sat $\ulcorner \alpha \leq \beta \urcorner$, then $\mathbf{N}$ sat ϕ. Assume the hypothesis. Thus, $u\mathbf{R}y$. Let A = the set of $\mathbf{R}$-least elements.

Let $T = \{a : a \in A \text{ and for some } z \text{ in } S, a\mathbf{R}z\}$.

By Lemma (2.6.3.2), $y = \sup_\mathbf{R} T$. Since $u \in \mathbf{U}$, for some $a \in A$, $a\mathbf{R}u$; and hence (by transitivity of $\mathbf{R}$) $a\mathbf{R}y$. By Definition (2.2.6) (3a), $a \in T$. Hence, for some $z \in \mathbf{U}$, z is $\mathbf{R}$-least in $\mathbf{U}$, $a\mathbf{R}z$, and $\langle \mathbf{R}, \mathbf{U}, \mathbf{A}_z^\alpha \rangle$ sat ϕ. By the condition (1) on models, $\inf_\mathbf{R}(u, z) \in \mathbf{U}$. Clearly, $\inf_\mathbf{R}(u, z) \mathbf{R} z$ and, since z is $\mathbf{R}$-least in $\mathbf{U}$, $\inf_\mathbf{R}(u, z) = z$. Similarly, $\inf_\mathbf{R}(u, z) = u$. Thus, $u = z$. Hence $\langle \mathbf{R}, \mathbf{U}, \mathbf{A}_u^\alpha \rangle$ sat ϕ; and, since β does not occur in ϕ, $\mathbf{N}$ sat ϕ.

Necessity: Show, if $\mathbf{N}$ sat ϕ then $\mathbf{N}$ sat $\ulcorner \alpha \leq \beta \urcorner$. Assuming the hypothesis, clearly $u \in S$. Hence, $u\mathbf{R}y$. The consequent follows.

In order to demonstrate the completeness of CIIII, it suffices to show that the model $\mathbf{M}$, which was constructed in the course of proving Lemma (2.5.4.14) with respect to CII, also satisfies the new conditions imposed on the models of CIIII, provided the axiom-scheme (AxS6) is added to the axioms of CII. In order to avoid undue repetition, we shall prove the analog of (2.5.4.14) under reference, by number, to the steps in the proof of (2.5.4.14).

In the proof of Lemma (2.5.4.14), let the steps (1)–(7) and (8a), (8b) stand as before. We insert, as (8c), the proof that the universe $\mathbf{U}$, characterized in (6), satisfies the condition (1) imposed by Definition (2.7.3.1) on models of CIIII; that is, we show that **U** is closed under infima of overlapping individuals. But this condition is common to models of CIII and to ones of CIIII. We have already shown, in connection with the completeness of CIII, that a universe satisfies this condition, provided that (Ax4) is a theorem of the system. Indeed, (Ax4) is also a theorem of CIIII, as we have noted in Section 2.7.2. Thus, step (8c) is identical to the proof given under the same number in connection with CIII.

Next, insert, unaltered, the proofs (9) and (10) occurring in (2.5.4.14), and continue:

(11) Show: for all formulas ϕ, **M** sat ϕ *in CII* if and only if $\phi \in K^*$. Thus, although the notions of a model and of satisfaction in the present proof are those appropriate to CIIII, we first have to show this subsidiary result (11) with respect to CII. However, since **M** is also a model of CII and since all theorems of CII are theorems of CIIII, the proof of (11) is the same as that of (10). It remains to be shown that **U** meets the condition (2) imposed on models of CIIII:

(12) Show: for all S, if S is non-empty and specifiable in **M** (relative to CII), then $\sup_\mathbf{R} S \in \mathbf{U}$.

Assume the hypothesis. Then, for some formula ϕ and for all x, $\langle \mathbf{R}, \mathbf{U}, \mathbf{A}_x^\alpha \rangle$ sat ϕ (in CII) if and only if $x \in S$.

For some z, $z \in S$; hence $z \in \mathbf{U}$; let $z = \{\gamma: \ulcorner At\, \gamma\, \&\, \gamma \leqslant \beta \urcorner \in K^*\}$. $z \neq 0$ and, by (7), $z = \mathbf{A}(\beta)$. $\langle \mathbf{R}, \mathbf{U}, \mathbf{A}_{A(\beta)}^\alpha \rangle$ sat ϕ. By Lemma (2.5.4.2), **M** sat ϕ_β^α (in CII). By (11), $\phi_\beta^\alpha \in K^*$. Also, since $z \neq 0$, for some γ, $\ulcorner At\, \gamma\, \&\, \gamma \leqslant \beta \urcorner \in K^*$.

Let $\psi(\gamma) = \ulcorner (\exists \delta) (\gamma \leqslant \delta\, \&\, \phi_\delta^\alpha) \urcorner$, for all γ.

Clearly, $\psi(\gamma) \in K^*$, and hence $\ulcorner (\exists \zeta)(At\, \zeta\, \&\, \psi(\zeta)) \urcorner \in K^*$. By (AxS6) and (1), there is a variable η, $\ulcorner (\forall \zeta)(At\, \zeta \to [\zeta \leqslant \eta \leftrightarrow \psi(\zeta)]) \urcorner \in K^*$. We show (12a): $\bigcup S = \{\zeta: \ulcorner At\, \zeta\, \&\, \zeta \leqslant \eta \urcorner \in K^*\}$ (which, clearly, $= \mathbf{A}(\eta)$).

Assume that $\zeta \in \bigcup S$, so that for some $y \in S$, $\zeta \in y$. Then $\langle \mathbf{R}, \mathbf{U}, \mathbf{A}_y^\alpha \rangle$ sat ϕ, y must be in **U**, and hence for some variable τ, $y = \mathbf{A}(\tau)$. By Lemma (2.5.4.2), $\langle \mathbf{R}, \mathbf{U}, \mathbf{A} \rangle$ sat ϕ_τ^α and, by (11), $\phi_\tau^\alpha \in K^*$. Since $\zeta \in \mathbf{A}(\tau)$, $\ulcorner At\, \zeta\, \&\, \zeta \leqslant \tau \urcorner \in K^*$. Hence $\psi(\zeta) \in K^*$. Also, $\ulcorner \zeta \circ \zeta \urcorner \in K^*$. Hence, $\ulcorner \zeta \leqslant \eta \urcorner \in K^*$, and $\zeta \in \mathbf{A}(\eta)$.

Conversely, assume that $\ulcorner At\, \zeta\, \&\, \zeta \leqslant \eta \urcorner \in K^*$. Hence $\psi(\zeta) \in K^*$. For some variable δ, $\ulcorner \zeta \leqslant \delta\, \&\, \phi_\delta^\alpha \urcorner \in K^*$. $\zeta \in \mathbf{A}(\delta)$. By (11), **M** sat ϕ_δ^α; hence $\langle \mathbf{R}, \mathbf{U}, \mathbf{A}_{A(\delta)}^\alpha \rangle$ sat ϕ. Thus, $\mathbf{A}(\delta) \in S$ and $\zeta \in \bigcup S$.

By (12a), $\bigcup S$ must be a member of **U**. Since the given relation **R** is that of inclusion, $\sup_\mathbf{R} S = \bigcup S$. This completes the proof of (12).

Next, it is obvious on the basis of (11) and the fact that **M** is a model of CIIII that for all formulas ϕ, **M** sat ϕ *in CIIII* if and only if $\phi \in K^*$. Thus, we obtain:

(13) For all formulas $\phi \in K$, **M** sat ϕ in CIIII.

We have shown the essential steps establishing the semantical completeness of CIIII.

2.8. NON-ATOMISTIC UNIVERSES OF INDIVIDUALS

Up to this point, we have restricted our discussion to that notion of an individual which was introduced by Goodman (1956), and which is analyzed in terms of a principle of individuation (2.3.3) and a principle of summation (2.3.8) which imply that every individual has atomic parts. Although Goodman, in 1956, seemed to regard this assumption of atomicity as central to the very notion of an individual, he has suggested in other contexts that individuals might conceivably lack atomic parts. Thus, in *The Structure of Appearance* he says: 'For all finite systems, and it is these that primarily concern us, there will be atomic individuals – individuals that have no others as systematic parts.'[23] The suggestion appears to be that there is a notion of an individual which does not primarily concern Goodman but to which the assumption of atomicity is not essential. Also, as we have observed, the calculus of individuals LGCI, which may be taken as one expression of Goodman's beliefs regarding individuals, does not imply the existence of atoms.

In Section 2.1, point (3), we have briefly examined the question whether it seemed reasonable to assume that every individual has atomic parts. We could not discover any compelling reasons for either accepting or rejecting this assumption. The question remains open, however, whether appropriate principles of individuation and of summation can be formulated if the assumption of atomicity is abandoned.

Recently, Yoes[24] has published an interesting though inconclusive discussion concerning principles of individuation which might be appropriate to individuals without atoms. He considered, in some detail, only one generating relation: the proper ancestral of membership (which x bears to y if x is connected by a finite 'upward' membership chain with y; see 1.3.1.52). However, with respect to this particular relation it is rather plausible to assume, not only in nominalistic systems, but also in set theory, that every item dominates some atom.[25] The interest of non-atomic systems seems far more striking if generating relations other than the ancestral of membership are contemplated. Examples of such relations will be given shortly. However, the difficulties, reported by Yoes, of formulating an appropriate principle of identification, may well raise doubts whether such principles can be stated with respect to part-whole relations in general. Happily, this can be done rather easily. Before doing so, let us return to the discussion of atomistic universes of individuals (Section 2.3) and seek to modify the notion of a universe in such a fashion that universes can, but need not be, atomistic.

We have earlier decided that a certain principle of summation is characteristic of atomistic universes; namely, that every individual in the universe is the sum of its atomic parts in that universe. It seems natural to formulate a principle of summation which is akin to this former one, but avoids reference to atoms. We introduce first, by definition, some intuitive notions which facilitate the reading of subsequent formulations, and then we shall inspect some candidates which might appear to be plausible principles of summation.

(2.8.1) DEFINITIONS: x is a *proper R-part of y* just in case R is a partial ordering, x bears R to y, and x differs from y; and x is a *proper R-part of y in U* if x is a proper R-part of y and x is in U.

Suppose that **R** is a part-whole relation (in the sense of 2.2.6), and that **U** is a subset of the field of **R**. Then one of the following principles may seem to be an appropriate condition on **U**, if **U** is to be a non-atomistic universe of individuals:

(2.8.2) For every x in **U** there exists a subset S of **U** such that $x = \sup_\mathbf{R} S$.

Informally: every individual (in **U**) is a sum of parts which are in **U**. However, since every individual is part of itself and the sum of its own singleton (see 2.2.2), the condition (2.8.2) is provable and imposes no restriction. We try next:

(2.8.3) For every x in **U** there exists a set S such that all members of S are proper **R**-parts of x in **U**, and $x = \sup_\mathbf{R} S$.

INDIVIDUALS

Informally: every individual (in **U**) is the sum of its proper **R**-parts in **U**. This condition, however, is too strong. We want to permit (though not require) that there exist atoms in **U**. And an atom, having no proper **R**-parts in **U**, cannot be the sum of its proper **R**-parts in **U**; that is:

(2.8.4) REMARK: If **U** is a subset of the field of a part-whole relation **R** and if x is **R**-least in **U**, then there is no set S of proper **R**-parts of x in **U** such that $x = \sup_\mathbf{R} S$.

For, 0 is not in the field of **R** and hence not in **U**. Hence, assuming the negation of (2.8.4), $\sup_\mathbf{R} S$ exists and is the same as $\sup_\mathbf{R} 0$. By definition and vacuously, $\sup_\mathbf{R} S$ bears **R** to everything. There are two distinct **R**-least elements, $\sup_\mathbf{R} 0$ bears **R** to each of them and is therefore identical with each of them: impossible.

But the following condition is neither trivial nor impossible:

(2.8.5) For every x in **U**, either x is **R**-least in **U** or else there exists a set S such that all members of S are proper **R**-parts of x in **U**, and $x = \sup_\mathbf{R} S$.

Informally: every individual (in **U**) is either atomic (relative to **U**), or else the sum of its proper **R**-parts (in **U**).

In order to intuit what is going on, we construct some examples. Let A be the set of all unit-sets of natural numbers; let R = the inclusion relation restricted to sets of numbers; and let U_1 = the set of all those sets x such that for some number **n**, x comprises all numbers except those less than **n**. That is,

$$U_1 = \{\{0, 1, 2, 3, \ldots\}, \{1, 2, 3, \ldots\}, \{2, 3, \ldots\}, \ldots\}.$$

Relative to inclusion, U_1 has no atomic members (since, for any number **n**, the set of all numbers excepting those less than (**n**+1) is properly included in the set of all numbers excepting those less than **n**). But U_1 does not satisfy the condition (2.8.5). For sums, in this example, are unions; and, whereas the set of all numbers is in U_1 and comprises 0, there is no union of other members of U_1 which could comprise 0. Thus, U_1 is atomless, but not a universe of individuals.

The result U_2 of throwing into U_1 all non-empty complementary sets of numbers is non-atomistic (since some sets fail to include atoms), though not atomless (since $\{0\}$ occurs). But U_2 is still not a universe of individuals satisfying (2.8.5) since, e.g., $\{0, 1\}$ is neither an atom nor a sum of proper subsets in U_2.

Next, let U_3 = the set of all those sets x such that for some number **n** and for some number **m** less than 2^n, x comprises all numbers which leave the remainder **m** upon division by 2^n. Thus, U_2 comprises the set of all numbers; the sets of all even and of all odd numbers; the four sets of all numbers which, upon division by 4 leave the remainders 0, 1, 2, or 3; and so forth. U_3 comprises infinitely many classes, each of which is infinite. Each set of numbers which, upon division by 2^n, leave the remainder **k** is the union of the following exclusive and proper subsets: (1) the set of all numbers which, upon division by 2^{n+1}, leave the remainder **k**, and (2) the set of all numbers which, upon division by 2^{n+1}, leave the remainder $(k+2^n)$. Thus, U_3 meets the condition (2.8.5) and qualifies as an atomless universe of individuals relative to inclusion.

For intuitive reasons which have been mentioned in Section 2.3, we refrain from imposing on all universes of individuals the condition that all sums of individuals shall be actualized. But we do find it plausible to accept the condition (2.8.5) and define:

(2.8.6) DEFINITION: **U** is a *(non-atomistic) universe of individuals for* **R** if and only if **R** is a part-whole relation, **U** is included in the field of **R**, and (2.8.5) for every x in **U**, either x is **R**-least in **U** or else there exists a set S such that all members of S are proper **R**-parts of x in **U**, and $x = \sup_R S$.

The following lemma is an analog of Lemma (2.3.12):

(2.8.7) LEMMA: Suppose that **R** is a part-whole relation and that **U** is included in the field of **R**. Then **U** is a (non-atomistic) universe of individuals for **R** if and only if the following condition is satisfied: for every x in **U** which is not **R**-least in **U** and for every y in the field of **R**: if every proper **R**-part of x in **U** bears **R** to y, then x itself bears **R** to y.

The proof is closely similar to that of Lemma (2.3.12). In showing the necessity, let S, in the earlier proof, be instead the set of proper **R**-parts of x in **U**.

As an immediate corollary of lemma (2.8.7), we obtain a principle of individuation which is appropriate to any universe of individuals:

(2.8.8) THEOREM: Suppose that **U** is a (non-atomistic) universe of individuals for **R**. Then, for all x and y in **U** which are not **R**-least in **U**: $x = y$ if and only if every proper **R**-part of x

INDIVIDUALS 77

in **U** bears **R** to y, and every proper **R**-part of y in **U** bears **R** to x.

We shall turn next to formal calculi whose theorems hold in non-atomistic universes of individuals.

2.9. THE NON-ATOMISTIC CALCULUS CI IV AND SOME EXTENSIONS

In Section 2.4 we observed that the Leonard-Goodman system LGCI is non-atomistic in the sense that it fails to imply that every individual has an atomic part. Upon deletion of its sum-axiom (AS6), LGCI may appear to have just about the right strength to hold in all non-atomic universes of individuals (as defined in 2.8.6). However, this is not the case. Indeed, the Leonard-Goodman system, with or without its sum-axiom, is unsound relative to non-atomistic universes, even if on such universes one imposes any of the closure conditions discussed in connection with atomistic universes.

In all systems so far considered, we have taken the notion of overlapping for primitive. This was done, partially, in order to facilitate comparisons with the calculus given by Goodman in *The Structure of Appearance*, where the same primitive is used; and partially because we shall later on interpret the calculus in terms which make 'overlapping' a more natural choice of a primitive than other known alternatives. In the original Leonard-Goodman system, the primitive 'is discrete from' is used instead; but it is clear that 'being discrete from' serves no better than 'overlapping'. However, 'overlapping', under its natural interpretation, is too weak to be taken as a primitive in non-atomistic systems.

To convince ourselves of this, consider the definition

(D1) $\quad x \leqslant y \leftrightarrow [x \circ x \,\&\, (\forall z)(z \circ x \to z \circ y)]$.

Given a non-atomistic universe **U** with respect to a part-whole relation **R**, and items x and y in **U**, the intended interpretation of the left-hand side of this biconditional is: x bears **R** to y; and that of the right-hand side is: there is a v in **U** such that $v\mathbf{R}x$ and for every z in **U**, if z and x have a common **R**-part in **U** then z and y have a common **R**-part in **U**. We shall construct a sample universe of which the right-hand side is true, but the left-hand side is false:

Let **U** be the family of those sets which comprise almost all natural numbers (that is, all of them, or all except a finite number of them). It is easily verified that **U** is an atomless universe of individuals with respect to inclusion. Let x be the set of all numbers, and y the set of all numbers except 0. Given any set z in **U**, there exists a set v in **U** which is included in both z and y. Hence, the right-hand side of (D1) is true. But its left-hand side is not, for x is not included in y. Furthermore, **U** is closed under sums and finite intersections. The decisive point however is that **U** is not closed under difference.

In the systems so far discussed, this difficulty arises specifically at the following places: In CII, the crucial Lemma (2.5.4.11), part (1), fails from left to right, given models whose universes are non-atomistic. In LGCI, the axiom (A4) is unsound in such models; but it is sound in models whose universes are closed under difference, particularly in ones which are full Boolean Algebras. However, the properties of complete Boolean Algebras are sufficiently known to justify our neglect of such systems, and to warrant attention to more general structures.

Instead, we shall formalize a system, CIIV, whose theorems hold in all universes of individuals and which, in this sense, is a most general theory generating analytic truths regarding individuals. Instead of 'overlapping', we shall choose the stronger primitive 'is part of'. As we have done in all systems, excepting LGCI, the symbol '=' of identity will be defined. This means, syntactically, that all logical axioms characteristic of identity in systems without existential import can be derived as theorems from the remaining axioms and a definition; and further, semantically, that statements of the form '$x = y$' will be true only of identical individuals, and not also of ones which enter an equivalence relation other than identity.

2.9.1. The Vocabulary and the Axioms of CIIV

The *vocabulary* of CIIV is that of CII. However, the unique primitive symbol of CIIV is the non-schematic two-place predicate '$\leqslant$'.

The *axioms* of CIIV are listed as follows: all logical axioms (L1)–(L3) remain as before (see Section 2.4). For all appropriate variables and terms, we have the following definitional axioms:

(Def. 1) $\quad \zeta < \eta \leftrightarrow (\zeta \leqslant \zeta \,\&\, \zeta \leqslant \eta \,\&\, \sim \eta \leqslant \zeta)$.

In words: ζ is a proper part of η just in case ζ is part of itself (or, ζ exists), ζ is part of η, but η is not part of ζ.

(Def. 2) $\quad \zeta = \eta \leftrightarrow (\zeta \leqslant \zeta \ \& \ \eta \leqslant \eta \ \& \ \zeta \leqslant \eta \ \& \ \eta \leqslant \zeta)$.

In words: ζ is identical with η just in case both ζ and η are parts of themselves (or, exist), and parts of one another.

(Def. 3) $\quad \zeta \circ \eta \leftrightarrow (\exists \alpha)(\alpha \leqslant \zeta \ \& \ \alpha \leqslant \eta)$.

In words: ζ overlaps η just in case ζ and η have a common part.

(Def. 4) $\quad At\zeta \leftrightarrow [\zeta \leqslant \zeta \ \& \ \sim (\exists \alpha) \alpha < \zeta]$.

In words: ζ is an atom just in case ζ is part of itself (or, exists) but has no proper part.

In addition, the following are proper axioms of CIIV:

(Axm. 1) $\quad [\zeta \leqslant \zeta \ \& \ (\forall \alpha) \phi] \to \phi_\zeta^\alpha$.

In words: If ζ is part of itself (or, exists) and if everything satisfies the condition ϕ, then so does ζ.

(Axm. 2) $\quad (\forall \alpha) \alpha \leqslant \alpha$.

In words: Every (actual) individual is part of itself (or, every actual individual exists).

(Axm. 3) $\quad (\zeta \leqslant \eta \ \& \ \eta \leqslant \tau) \to \zeta \leqslant \tau$.

In words: If ζ is part of η and η is part of τ, then ζ is part of τ. '$\leqslant$' is transitive.

(Axm. 4) $\quad [(\exists \alpha) \alpha < \zeta \ \& \ (\forall \alpha)(\alpha < \zeta \to \alpha \leqslant \eta)] \to \zeta \leqslant \eta$.

In words: If ζ has a proper part and if every proper part of ζ is part of η, then ζ itself is part of η. This is the syntactical counterpart of the condition which, according to lemma (2.8.7), is characteristic of universes of individuals.

Due to (Axm. 4) and the definitions, as well as (Axm. 2), it is provable that an individual is part of itself just in case it overlaps itself; and also that everything overlaps itself. Beyond that, briefly, the *theorems* of CIIV are just those theorems of CII which do not imply the existence of atoms. These include all theorems listed in Section 2.5.2, except 27, 28, 30, 31, 32, 33, 34, and possibly 35.

2.9.2. *The Semantics of CIIV*

The theorems of CIIV are intended to be necessarily true of individuals in all universes. Hence, the universes of models in CIIV should be universes of individuals in the sense of (2.8.6):

(2.9.2.1) DEFINITION: **M** is a *model* of CIIV if and only if there exist **R**, **U**, and **A** such that $\mathbf{M} = \langle \mathbf{R}, \mathbf{U}, \mathbf{A} \rangle$ and
(1) **U** is a universe of individuals for **R**, and
(2) **A** is a function whose domain is a set of variables and whose range is included in **U**.

The notion of *satisfaction*, **M** sat ϕ, is defined for all models **M** and for all primitive formulas ϕ in the customary manner made explicit in (2.5.3.3), except that the primitive atomic formulas now have the form $\ulcorner \alpha \leqslant \beta \urcorner$ and the appropriate satisfaction condition (replacing (1) in 2.5.3.3) will read:

(1′)   **M** sat $\ulcorner \alpha \leqslant \beta \urcorner$ just in case $\mathbf{A}(\alpha)$ bears **R** to $\mathbf{A}(\beta)$.

(2.9.2.2) REMARK: If **M** sat $\ulcorner \alpha \leqslant \beta \urcorner$, then both α and β are in the domain of **A** and both $\mathbf{A}(\alpha)$ and $\mathbf{A}(\beta)$ are in the universe of **M**.

All other semantical notions of CIIV are precisely like those of CII.

The theorems of CIIV are just those formulas which are satisfied by all models of CIIV. Brief outlines of proofs which support this contention are given in the next section.

2.9.3. *The Semantical Adequacy of CIIV**

All the lemmas (2.5.4.1)–(2.5.4.12) stated in connection with CII, continue to hold in CIIV. In addition, if $\mathbf{M} = \langle \mathbf{R}, \mathbf{U}, \mathbf{A} \rangle$ is a model, then **M** sat $\ulcorner \alpha \circ \beta \urcorner$ just in case there exists an x in **U** such that x bears **R** to both $\mathbf{A}(\alpha)$ and $\mathbf{A}(\beta)$.

Given these lemmas, the semantical soundness of CIIV is easily verified.

INDIVIDUALS

It is tedious and not particularly difficult to modify the completeness proof of CII so that it applies to CIIV. For this reason, we only give a few hints.

In the proof of lemma (2.5.4.14), let the steps (1)–(5) be as before. Then,

(6) Let $\mathbf{U}=$ the set of all items x such that $x \neq 0$ and for some variable α, either (a) ⌜$At\ \alpha$⌝$\in K^*$ and $x=\{\beta:$ ⌜$\beta=\alpha$⌝$\in K^*\}$, or (b) ⌜$At\ \alpha$⌝$\notin K^*$ and $x=\{\beta:$ ⌜$\beta<\alpha$⌝$\in K^*\}$.

(7) Let $\mathbf{A}=$ that function whose domain is the set of all variables such that ⌜$\alpha \leqslant \alpha$⌝$\in K^*$ and for each such α, (a) if ⌜$At\ \alpha$⌝$\in K^*$ then $\mathbf{A}(\alpha)=\{\beta:$ ⌜$\beta=\alpha$⌝$\in K^*\}$, and (b) if ⌜$At\ \alpha$⌝$\notin K^*$ then $\mathbf{A}(\alpha)=\{\beta:$ ⌜$\beta<\alpha$⌝$\in K^*\}$.

Then proceed much as in the previous proof, but reasoning by cases, and replacing references to atoms by references to proper parts.

2.9.4. *Extensions of CIIV*

It is rather clear that the non-atomistic calculus CIIV admits of extensions similar to the extensions CIII and CIIII of CII.

For example, if $\mathbf{M}$ is a model of CIIV, let us define the notion of a *specifiable* set relative to CIIV exactly as we did in Definition (2.7.1). Let a new system, say CIV, have for its models all those models of CIIV which satisfy, in addition, the closure conditions (1) and (2) (relative to CIIV) stated in Definition (2.7.3.1). That is, roughly, let the universes of the new models be closed under non-empty products and under sums of non-empty specifiable subsets.

On the syntactical side, let the axiom set of CIIV be augmented by axioms of the form:

(AS6) $\quad (\exists \alpha)\ \phi \rightarrow (\exists \gamma)\ (\forall \beta)\ [\beta \circ \gamma \leftrightarrow (\exists \alpha)\ (\phi\ \&\ \beta \circ \alpha)],$

which is the sum-axiom of LGCI.

It is true, but will not be shown here, that CIV is semantically adequate.

2.10 SEQUENTIAL INDIVIDUALS AND THE CALCULUS OF INDIVIDUAL RELATIONS

As Hempel[26] has pointed out, we seem to have a reasonably clear notion of individuals which differ just with respect to the sequential arrangement among their individual parts and which appear to differ, therefore, even when they have the same content. This conception of individuals conflicts with Goodman's. Goodman[27] claimed that, leaving aside mathematical sophistications and set theoretic constructions, he could not find any everyday instances of distinct things which are composed of the same things.

Now, in as much as Goodman appeals to everyday instances of things, rather than to a philosophically refined notion of an individual, his inability to discover such things is somewhat surprising. Whenever people speak of cloth patterns, flower arrangements, fleet formations, planetary systems, stellar constellations, clusters of berries, sentences, organisms, and so forth, they are naturally taken to speak of things which can differ just with respect to the arrangement among their parts.

Furthermore, in their efforts to characterize a syntax adequate for expressing arithmetic, Goodman and Quine[28] find it necessary to introduce, as a primitive three-place predicate, the symbol 'C', where '$Cxyz$' is intended to be true of individuals x, y, and z just in case x is the concatenate of (the result of writing in order from left to right) the individual characters y and z. Although no precise interpretation of this primitive has been given, it must be assumed that some semantical treatment of concatenates is available which does not have the consequence that the concatenate x of two individual inscriptions y and z is some entity other than an individual. Since the concatenate of two characters is determined by the order in which they occur, it would seem that Goodman is compelled to take account of individuals whose parts are sequentially ordered.

If two individuals x and y are 'basic' in the sense that they are not themselves concatenates, then the result of juxtaposing them has essentially the same properties as the ordered pair (x, y). It is essential to ordered pairs that they are distinct whenever the order or arrangement among their constituents differs. This property of ordered pairs is formally expressed by the condition:

(1)   if $(x, y) = (u, v)$, then $x = u$ and $y = v$.

Of this it is an immediate consequence that

(2)   if $x \neq y$, then $(x, y) \neq (y, x)$.

In employing the notion of an ordered pair in the present context, we shall not assume that ordered pairs are defined in any preconceived manner (as, e.g., in 1.3.1.17). We merely suppose that the expression 'the ordered pair of... and ---' refers to some operation which, when applied to individuals, does not result in a non-individual, and regarding which the condition (1) always holds. Since pairs of individuals and concatenates of basic characters share all properties which presently concern us, we shall use the notation '(x, y)', in the present context, indiscriminately to refer to pairs or to such concatenates. Generally, we employ the notation '$(x_1,..., x_n)$' to refer to any sequential individuals, so long as they always satisfy the law:

(3)   If $(x_1, ..., x_n) = (y_1, ..., y_m)$, then $n = m$ and for all $i \leq n$, $x_i = y_i$.

Prima facie, the notion of an ordered pair which is also an individual appears to be contradictory. For, assume that x and y are any distinct atoms and suppose that (x, y) and (y, x) are new individuals whose only atomic parts are x and y. Then, by Goodman's principle of individuation (2.1.1), it must be the case that $(x, y) = (y, x)$, contrary to the law (2). Intuitively, if individuals differ only if they have different parts, then entities which have exactly the same parts and differ only with respect to the arrangement of these parts, cannot be individuals. In particular, it may seem impossible to endow such formulas as '$Cxyz$' with a semantical interpretation which ever makes them true of individuals. And if such doubts arise, they should also be extended to the Goodman-Quine program of expressing arithmetic in a language acceptable to nominalists.

However, at least four different methods come to mind which would permit us to construe ordered pairs as individuals. Under the assumption that x and y are distinct atoms, these four approaches can be roughly outlined as follows:

(A) We grant that x and y are the only parts of both (x, y) and (y, x), but insist that (x, y) and (y, x) are wholes of these parts with respect to different part-whole relations.

(B) We concede that x and y are the only parts of (x, y) and (y, x), we admit only one part-whole relation, we insist that the values of the pairing operation shall not be entities other than individuals, and yet we still embrace, without contradiction, the conclusion that at most one of them is an individual.

(C) We deny that x and y are the only parts of (x, y) and (y, x) and insist that there exist other parts, which do not meet the eye, and whose occurrence somehow serves to mark the positions occupied by the two evident parts x and y.

(D) We construe (x, y) and (y, x) as having parts other than x and y. However, the function of these extra parts is not conceived to be that of marking positions, but rather that of linking x to y and y to x.

Let us consider each of these proposals in turn and outline some formal developments of those approaches which seem more promising.

2.10.1. *Order Implied by Different Part-Whole Relations*

Imagine a child using the same building blocks to construct first a tower and later a wall. In this case, the blocks enter different arrangements and form different wholes at different times. Although, in this description, we have taken account of 'temporal divisibility' (to use Goodman's expression [29]), we did not feel compelled to assume a temporal divisibility *of the blocks*. That is to say, although we admit that the blocks form part of a tower and part of a wall at different times, there is no need to admit that there are temporally different parts of the blocks which enter these two constructions. One could, of course, maintain that each building block 'really' consists of temporally discrete slices and that the earlier block-stage which is part of the tower differs from but is continuous with the later block-stage which is part of the wall. But it is far more natural to describe the situation by saying that the very same blocks form part of a tower and later also of a wall. Intuitively, it is not the parts which are different in the two constructions, but rather the mode of composition: the manner in which wholes are formed from these parts. What is meant by different modes of composition, in this context, is just what we mean by different part-whole relations. Hence, the point is made that from the very same building blocks different individual wholes may be formed relative to different part-whole relations. In particular, ordered pairs may be regarded as individual wholes of their constituents with respect to different part-whole relations.

In spite of the initial plausibility of this approach, it does not appear to lend itself easily to systematic treatment. Consider how one would deal with the simplest case, that of ordered pairs:

One would seek to characterize two part-whole relations **R** and **R'** in such a fashion that the pair (x, y) is the sum of x and y relative to **R** and (y, x) is the sum of x and y relative to **R'**. Suppose that **R** and **R'** were so contrived that they have the same atoms. We would next replace our notion 'being a universe of individuals for **R**' by the notion 'being a universe of individuals for both **R** and **R'**'. If (x, y) and (y, x) are to come out distinct, an appropriate principle of individuation on universes in this new sense should presumably read: $x = y$ just in case for all atoms z, z bears both **R** and **R'** to x if and only if z bears both **R** and **R'** to y. Individuals in a universe should be sums of their atoms simultaneously with respect to **R** and **R'**. But neither x nor y nor any parts that come to mind bear both relations to their pairs.

Even if the details of some such approach could be worked out, it would not seem to accord with the spirit of nominalism. For once we admit individuals which differ with respect to different modes of composition, the door is open for admitting as individuals the results of many set theoretic operations which a nominalist could not endorse.

2.10.2. *Order Implied by the Existence of Certain Composites*

Let us return to the consideration of just one part-whole relation. If x and y are atoms, we grant that no atoms other than x and y are part of the ordered pairs (x, y) and (y, x). From this it follows, as we have seen, that at most one of the given pairs can be an individual. However, this conclusion does not imply that one of the pairs is a non-individual, in the sense that one of them exists and fails to be an individual. It merely follows that one of the two pairs cannot exist.

In the case of concatenates of distinct basic characters x and y, this conclusion is quite intuitive. After all, if x and y are particular distinct inscriptions, then at most one of the concatenates (x, y) or (y, x) can be concretely actualized in a specific region of a piece of paper. Thus, if '$Cxyz$' means 'x is the concatenate of y and z', it should be a theorem that

(2.10.2.1) $\quad (\exists x)\, Cxyz \rightarrow \sim (\exists x)\, Cxzy.$

That is, if the concatenate of the inscriptions y and z exists, then the concatenate of z and y does not. Furthermore, if '$x + y$' denotes the individual sum of x and y, it should be provable that

(2.10.2.2) $\quad Cxyz \rightarrow x = y + z,$

that is, the concatenate of two characters y and z, if it exists, is just their sum. For the concatenate of two items, by our present agreement, has precisely the same parts as their sum.

If, instead of a three-place predicate '$Cxyz$' meaning 'x is the concatenate of y and z', one employs identity and a two-place operation expression to say by '$x=(y,z)$' that x is the same as the concatenate of y and z, it may not be clear how the analog of (2.10.2.1) can be expressed. Indeed, in currently favored systems of logic, operation expressions, such as '(x, y)', are so construed that the result of applying the operation in question to the items x and y must always exist. This convention may account for the fact that it seems contradictory to permit mention of the concatenates (x, y) and (y, x), while denying that both of them exist. However, alternative systems of logic are available.

The logic underlying our calculi of individuals was unconventional in that variables without values were admitted. It is also possible to admit non-denoting compound terms; in particular, ones which are formed by applying operation expressions to other terms.[30] If τ is any term, then formulas of the form '$\tau \circ \tau$' can be used, in calculi of individuals, to express that τ denotes some individual, and the formulas can be read: 'τ exists'. Let us extend a calculus of individuals by admitting operational terms of the form '(x, y)', to be read 'the concatenate of x and y'. Then the analog of (2.10.2.1) can be expressed as follows:

(2.10.2.3) $(x, y) \circ (x, y) \rightarrow \sim (y, x) \circ (y, x)$.

While the introduction of appropriate symbolisms such as Goodman's and Quine's primitive 'C' or our '(x, y)', may give the illusion of understanding individual concatenates, a genuine understanding can hardly be claimed unless we provide a precise semantical interpretation of the symbolism. Such an interpretation may be given by specifying under what conditions formulas of the form '$Cxyz$' shall be satisfied in the models of a calculus of individuals, or what, if anything, may be denoted by the term '(x, y)' in such models. Roughly, an interpretation of '(x, y)' may take the following form:

(2.10.2.4) '(x, y)' denotes an individual z in the model M if and only if z is the sum of x and y and, in addition, ϕ,

where ϕ is some condition which serves to distinguish those sums which

INDIVIDUALS

are also concatenates from other sums. We shall not provide a complete and carefully thought out theory of this sort, but only a rough outline of a semantical interpretation of concatenates which seems reasonable.

Take the models of CIIII (in whose universe all sums exist), and add to them a new auxiliary constituent **L**, so that our new models have the form $\langle \mathbf{R}, \mathbf{U}, \mathbf{A}, \mathbf{L} \rangle$, where $\langle \mathbf{R}, \mathbf{U}, \mathbf{A} \rangle$ is a model in the sense of Definition (2.7.3.1). **L** shall be a binary relation among objects in the universe **U**. Informally, '$x\mathbf{L}y$' may be read 'x is immediately to the left of y'. Intuitively, whenever x is immediately to the left of y, the concatenate of x and y should exist and be just the sum of x and y. Hence, the combination '$x\mathbf{L}y$ and $\sup_\mathbf{R}\{x, y\}$ is so-and-so' may be read: 'the concatenate of x and y exists and is so-and-so'. We impose on members of the universe **U** of a model the following conditions:

(1) For every non-empty subset S of **U** there exists an element x of S such that no element y of S bears **L** to x.

Informally: every non-empty set of individuals has a left-most element. (1) asserts that **L** is a well-founded relation in **U**. It has the following implications: (a) Nothing is immediately to the left of itself (**L** is irreflexive); (b) If x is immediately to the left of y, then y is not immediately to the left of x (**L** is asymmetric); (c) It is not the case that x is immediately to the left of y, y of z, and z of x (there are no three-term **L**-cycles); in fact, (d) there are no **L**-cycles of any length; also (e) there are no infinite chains of individuals extending to the left; and finally (f) we have the principle of induction: if K is any subset of **U** and if x is in K whenever all individuals to the left of x are in K, then all individuals are in K.

(2) If x, y, and z are **R**-least in **U**, then
 (a) if $x\mathbf{L}z$ and $y\mathbf{L}z$, then $x = y$, and
 (b) if $x\mathbf{L}y$ and $x\mathbf{L}z$, then $y = z$.

In words: no two atoms are immediately to the left or immediately to the right of any given atom. That is, **L** is linear among atoms. We hesitate, however, to require that **L** shall be linear among sums of atoms.

(3) $y\mathbf{L}z$ and $x\mathbf{L}\sup_\mathbf{R}\{y, z\}$ if and only if $x\mathbf{L}y$ and $\sup_\mathbf{R}\{x, y\}\mathbf{L}z$.

Informally: the concatenate of y and z exists and x is immediately to the left of it just in case the concatenate of x and y exists and is immediately to the left of z.

Since concatenates like (x, y), if they exist, are just the sum of x and y, the associative law of concatenation $(x,(y, z))=((x, y), z)$ will hold by virtue of the corresponding law governing sums, provided the existence of the right-hand concatenates is guaranteed given the existence of the concatenates on the left, and vice versa. The existence of these concatenates is assured by the condition (3).

(4)   If $x\mathbf{L}y$ and $y\mathbf{L}z$, then it is not the case that $x\mathbf{L}z$.

In words: if x is immediately to the left of y and y is immediately to the left of z, then x is not immediately to the left of z; that is, $\mathbf{L}$ is intransitive. This is what is meant by 'immediately' in 'immediately to the left'.

(5)   If $x\mathbf{L}y$ and $y\mathbf{L}z$, then $x\ \mathbf{L}\ \sup_{\mathbf{R}}\{y, z\}$.

Informally: if x is immediately to the left of y, and y of z, then x is immediately to the left of the concatenate of y and z. It follows, with (3), that the concatenate of x and y is also to the left of z.

From (4) and (5) it follows that $\mathbf{L}$ is intransitive for any finite number of terms. For example, if $x\mathbf{L}y$ and $y\mathbf{L}z$ and $z\mathbf{L}u$, then it is not the case that $x\mathbf{L}u$.

(7)   If $a\mathbf{L}b$, $x\mathbf{L}y$, and $\sup_{\mathbf{R}}\{a, b\}=\sup_{\mathbf{R}}\{x, y\}$, then either (i) $a=x$ and $b=y$, or (ii) for some e in $\mathbf{U}$, $x\mathbf{L}e$, $e\mathbf{L}b$, $a=\sup_{\mathbf{R}}\{x, e\}$ and $y=\sup_{\mathbf{R}}\{e, b\}$, or (iii) for some e in $\mathbf{U}$, $e\mathbf{L}y$, $a\mathbf{L}e$, $b=\sup_{\mathbf{R}}\{e, y\}$, and $x=\sup_{\mathbf{R}}\{a, e\}$.

This condition is a reformulation of the familiar principle that concatenates are identical only if they are either identical term for term or else instances of the associative law governing concatenates.

Let us say that an item x is *basic* if there exist no y and z such that $y\mathbf{L}z$ and $x=\sup_{\mathbf{R}}\{y, z\}$. Thus, basic individuals are ones which are not concatenates. Evidently, all atoms are basic; but certain sums which are not linear strings may also be basic. Then the following principle of induction would also appear suitable:

(8)   For every subset S of $\mathbf{U}$: if all basic elements are in S, and if $\sup_{\mathbf{R}}\{x, y\}$ is in S whenever both x and y are in S and $x\mathbf{L}y$, then for all x and y, if $x\mathbf{L}y$ then $\sup_{\mathbf{R}}\{x, y\}$ is in S.

If a model $\mathbf{M}=\langle \mathbf{R}, \mathbf{U}, \mathbf{A}, \mathbf{L}\rangle$ satisfies such conditions, we shall say that a variable α *denotes* x in **M** just in case $x = \mathbf{A}(\alpha)$ (which implies that α is in the domain of the assignment **A** and that x is in **U**); and we add that a compound term of the form $\ulcorner(\zeta, \eta)\urcorner$ *denotes* x in **M** if there exist y and z such that ζ denotes y in **M**, η denotes z in **M**, y bears **L** to z, and $x = \sup_{\mathbf{R}}\{y, z\}$. Thereupon we can say that **M** *satisfies* formulas of the form $\ulcorner\zeta \circ \eta\urcorner$ just in case for some x and y, ζ denotes x in **M**, η denotes y in **M**, and for some z in **U**, z bears **R** to both x and y.

This rough sketch of proposals which are advanced only very tentatively may at least serve to convince us that the basic approach under discussion seems workable.

Although it looks as if a coherent treatment of sequential individuals along these lines is feasible, the present approach is disadvantageous in some respects. To begin with, we had to rely on an auxiliary relation **L** on models. If it were possible (and we shall see that it is) to characterize sequential individuals solely in terms of a part-whole relation, such economy would seem preferable.

Furthermore, there are occasions where not only those sequences of individuals are needed which enter some specific characterizable relation, but where one would like to have sequential individuals which exist under all permutations of their parts. For example, it is tempting (and hard to avoid) to construe relations as sums of sequential individuals just as, in set theory, one defines relations as sets of sequences. Leaving aside those difficulties which may arise from the differences between sums and sets, the problem remains that such a program requires the existence of sequences whose parts are arranged in various orders. For example, if the relations 'is taller than' and 'is a father of' are to be totalities of pairs, then it can happen that a pair (x, y) must exist in order to enter the one relation, and the pair (y, x) must also exist if it is to form part of the totality constituting the other relation. But our present approach, according to which x and y may be the only parts of both pairs, precludes that both of them are existing individuals.

2.10.3. *Order Implied by Positional Parts*

Suppose we decide that the atoms x and y shall be parts of the ordered pairs (x, y) and (y, x), and that both of these pairs shall be actualized in some given universe of individuals. Then, as we have seen, the principle of individuation implies that at least one of these pairs must have parts other than x and y. Additional parts which come to mind are ones whose only role is that of indicating the positions of x and y.

Goodman's treatment of 'concreta'[31] illustrates the use of such special parts which, in his system of qualia, are regarded as phenomenal places and instants. 'Minimal concreta', in that system, are sums having for qualitative parts some phenomenon of the expected sort, such as a particular shade of color, and in addition position qualia which serve to assign appropriate locations to the other qualia. The 'togetherness' of such qualia in a minimal concretum is indicated by a special primitive predicate 'W' (where 'Wxy' is read 'x is with y'). It is essential of this relation that only discrete individuals (ones which have no common qualitative parts) are 'together'; and that whenever two individuals are 'together', so are any two discrete parts of their sum. For example, if we simplify Goodman's account by leaving aside temporal considerations and by distinguishing only the two positions 'left' and 'right', then the object (1) which is blue on the left and yellow on the right would be that individual whose atomic qualitative parts are 'left', 'right', 'blue' and 'yellow', and such that 'blue' is together with 'left' and 'yellow' is together with 'right'. By contrast, the individual (2) which is yellow on the left and blue on the right would have the same atomic parts, but would be such that 'yellow' is together with 'left' and 'blue' with 'right'. However, since the distinct items (1) and (2) would have the same atomic parts, they cannot both exist (we would have to add '...at the same time', if times were taken into account).

It is of some interest to note that in a calculus of individuals, such as CII, which is not closed under all sums of individuals, the togetherness-relation can be reduced to the part-whole relation. For if the atoms 'left', 'right', 'blue' and 'yellow' and the sum of them all is in a universe of individuals, it might be stipulated that either (1) the sum of 'left' and 'blue' and the sum of 'right' and 'yellow' shall also occur in the same universe, but the sums of 'left' and 'yellow' and of 'right' and 'blue' shall not (in which case the total sum would be regarded as an individual which is blue on the left and yellow on the right); or (2) only the sum of 'left' and 'yellow' and that of 'right 'and 'blue' shall be in the same universe (which would make the total sum an individual which is yellow on the left and blue on the right). If times are taken into account, then an individual which is blue on the left and yellow on the right at t_1, but yellow on the left and blue on the right at t_2, would be characterized by the fact that it breaks down, in a given universe, into the sums of blue, left, and t_1; yellow, right, and t_1; yellow, left and t_2; and blue, right, and t_2. In this fashion, the 'togetherness' of parts can be explained in terms of the existence of certain sums of parts; and concreta are distinguished by the manner in which they break down into partial sums. This avenue is, however, not open to us in systems, such as Goodman's, which imply that all sums of individuals exist. It remains problematic, in such systems, how Goodman's primitive 'W' (or some other relation connecting non-positional parts with positional ones) could receive a precise and intuitive semantical interpretation.

The approach outlined so far does not appear more advantageous than that discussed in the previous section. Since ordered pairs like (blue, yellow), (yellow, blue) are still construed as having the same atomic parts 'blue', 'yellow', 'left', 'right' (or other positions), they cannot both be individuals (at the same time, if we add times). If a special primitive like 'togetherness' has to be employed anyway, why could we not dispense with positional parts altogether and just rely on the method outlined in the preceding section of introducing some primitive like 'is immediately to the left of' to say that 'blue' is immediately to the left of 'yellow' or vice versa? If more complex positional relations are desired, some primitive like 'is in the spatio-temporal neighborhood of' might be employed. Not only does the present approach still fail to give us both of the sequential individuals (blue, yellow), (yellow, blue) (at the same time), but it has decided disadvantages over the previous method of construing ordered pairs. For it compels us to endorse as individuals certain positional parts which are intuitively abstract. Hence, if we should wish to restrict universes of discourse to individuals all of whose parts are in some sense concrete, a reference to parts whose only function is that of marking positions would seem to lead outside the universe of discourse.

If it is desired that the individuals (x, y) and (y, x) shall both exist, then it is not enough to endow each pair with additional parts indicative of positions. For instance, if each of the pairs has just the parts 'left', 'right', x, and y, then they are identical even though we have thrown in additional parts 'left' and 'right'. If pairs are to be distinguished, they must be construed as having *different* positional parts. In rough outline, this may be achieved as follows:

Designate any two **R**-least elements of a part-whole relation **R** by the names 'l' and 'r' for 'left' and 'right'. Thus, any atoms whatever may serve as positional parts. Consider (and definitionally delimit) just those universes of individuals with respect to **R** which do *not* comprise the items l and r as individuals. If **U** is such a universe, each of its atoms shall comprise one of the items l or r. The items l and r are therefore construed as **R**-least elements (relative to the entire field of 'possible' individuals), but not as **R**-least elements of the universe (of 'actualized individuals') **U**. Then the ordered pair (x, y) can be construed as consisting of two atoms of **U**: (1) the sum of x and l, and (2) the sum of y and r, both sums being irreducible within **U**. On the other hand, the pair (y, x) may consist of the sum of y and l and the sum of x and r, both of these sums being atomic relative to the universe **U**. Thus, within the given universe, (x, y) and (y, x) would have no parts in common and would therefore be distinct. For our principle of individuation identifies only those objects which have the same parts relative to the given universe.

On this last approach, positional parts have been so employed that both of the pairs (x, y) and (y, x) can be regarded as existing distinct individuals. But the price had to be paid of admitting positional parts l and r which are never in the universe of discourse, and to which, under customary interpretations, no expression of the object-language can make reference. For this reason, it seems doubtful that such a theory can be adequately expressed in any object-language acceptable to a nominalist. The admission of positional parts does not look like a very promising approach.

2.10.4. *Order Implied by Relational Individuals and their Calculus*

The paradigm cases of entities which a nominalist would allow appear to be concrete macroscopic objects. One can distinguish a left and a right side of each such object (any two sides, no matter how distinguished, will do). If such an object is also regarded as atomic (that is, if any parts it may have are either no longer concrete or no longer macroscopic), then its two sides can not be construed as proper parts in the same universe of discourse. Still, we can think of such an object as an ordered pair

(x, y) whose left side x and right side y are not themselves individuals in the given sense. In particular, if the two sides of an atom are the same, it can be conceived as a pair (x, x) whose left and right sides are distinguishable only by their positions. Let us call such uniform atoms 'units'. Given two units (x, x) and (y, y), we can construe as a new atom the object (x, y) whose left side is thought of as the right of (x, x) and whose right side is regarded as the left of (y, y). We call this intermediate atom a 'link', respectively, between (x, x) and (y, y). Now, the concatenate of the units (x, x) and (y, y) can be thought of as comprising three atoms: the units (x, x) and (y, y) and, in addition, the link (x, y). Similarly, the concatenate of the respective items (y, y) and (x, x) will contain the additional atom (y, x). Since the links (x, y) and (y, x) are distinct, the concatenates of the given units, taken in opposite order, will have different atomic parts, and will therefore be distinct. It seems that sequential individuals can exist which differ only with respect to the arrangement among their units, provided we endorse as new individuals all links between those units. However, it remains to be seen whether an overall adequate theory regarding such individuals can be constructed. To this end, we turn first to semantical considerations.

We want to characterize certain part-whole relations whose atoms shall essentially be ordered pairs. By the definition (2.2.6) of 'part-whole relation', all sums of such pairs must exist. Our paradigm part-whole relations so far have been inclusion relations confined to some set, and our typical sums have been unions. Given such relations, it is more intuitive to consider sets of ordered pairs (that is relations) rather than ordered pairs themselves. Let us first explain what is meant by a part-whole relation which is also an inclusion relation among relations:

(2.10.4.1) DEFINITION: **R** is an *IR-relation* (a relation of inclusion among relations) just in case there exist **P**, **A**, and **F** such that **P** is an infinite set, **A** is the set of all unit-sets of pairs whose constituents are in **P**, **F** is the set of all non-empty unions of elements of **A**, and **R** is the inclusion relation restricted to **F**.

Let us call a universe of individuals relative to an IR-relation a 'universe of individual relations', since the individuals in such a universe are, in fact, relations. Given a universe **U** of individual relations, the least individuals occurring in **U** need, of course, not be singletons of pairs, but may be any non-empty relations. Among such atomic relations, we want to distinguish links and units in a fashion which roughly corresponds to the distinction made earlier with respect to pairs viewed as the two sides of concrete things.

A link between two relations x and y in a universe **U** is conveniently taken to be the relative product of their converses (see 1.3.1.40, 1.3.1.23). For example, if $\{(a, b)(c, d)\}$ and $\{(v, x), (y, z)\}$ should be two atoms in **U**, then their respective link should be the set $\{(b, v), (b, y), (d, v), (d, y)\}$ which relates the last terms of the first set with each of the first terms of the second set.

It is less clear which atoms in **U** should be regarded as units. In analogy to what we said earlier, it would seem natural to regard as units just the identity relations confined to some set (see 1.3.1.36). For example, if $\{(x, x), (y, y)\}$ should be an atom in **U**, it looks like a suitable unit. However, in view of an axiomatic treatment yet to be given, we want it to be the case that all atoms are also links and that, in particular, every unit is linked by itself to itself. Recalling our earlier notion of links and units as pairs, it is clear that the unit (x, x) is linked by (x, x) to (x, x). This property is not forthcoming for the set $\{(x, x), (y, y)\}$. So, we add to it the requisite links, and regard instead the set $\{(x, x), (y, y), (x, y), (y, x)\}$ as an appropriate unit. We can formally define these notions as follows:

(2.10.4.2) DEFINITIONS: Suppose that **R** is an IR-relation and that **U** is an atomistic universe of individuals for **R** (in the sense of 2.3.11). Then
(1) x is a *unit* in **U** just in case x is **R**-least in **U** and for all a and b, (a, b) is in x if and only if both (a, a) and (b, b) are in x; and
(2) x is a *link* between y and z just in case y and z are units in **U** and $x=$ the set of all pairs (a, b) such that (a, a) is in y and (b, b) is in z.

We impose on universes of individual relations the twin requirements that (1) for any two units in **U** there shall exist an atomic link between them in **U**, and (2) every atom in **U** shall link two items in **U**. The second requirement has the effect that every atomic relation in **U** is either a unit (which links itself to itself), or proper link. We define:

INDIVIDUALS 95

(2.10.4.3) DEFINITION: **U** is a *universe of individual relations* for **R** just in case **R** is an IR-relation, **U** is an atomistic universe of individuals for **R**, and
  (1) for all x and y: if x and y are units in **U**, then there exists an **R**-least element z of **U** such that z is a link between x and y, and
  (2) for every **R**-least element x of **U** there exist y and z such that x is a link between y and z.

Let the *models* of the present theory of individual relations have the form $\mathbf{M} = \langle \mathbf{R}, \mathbf{U}, \mathbf{A} \rangle$ where **M** is a model of CII and **U** is, in addition, a universe of individual relations. Let the language of CII be enriched by an additional non-schematic three-place predicate 'L', where formulas of the form '$Lxyz$' are read 'y links x to z'. The satisfaction conditions for our new formulas shall be just those of CII (see 2.5.3.3), under addition of the following clause designed for atomic formulas of the new sort:

(2.10.4.4) If $\mathbf{M} = \langle \mathbf{R}, \mathbf{U}, \mathbf{A} \rangle$ is a model, then **M** *satisfies* $\ulcorner L\alpha\beta\gamma \urcorner$ just in case $\mathbf{A}(\beta)$ is a link between $A(\alpha)$ and $A(\gamma)$, and all three, $A(\alpha)$, $A(\beta)$, and $A(\gamma)$ are **R**-least in **U**.

Thus, 'L' serves to express linking relations among atoms of **U**.

The *axioms* of the theory of individual relations shall be those of CII, under addition of the following axioms peculiar to the relation 'L':

(L1) $(Lxyz \ \& \ Layb) \rightarrow (x=a \ \& \ z=b)$.

In words: an individual y links at most two items.

(L2) $Atx \rightarrow (\exists y)(\exists z) Lyxz$.

In words: every atom is a link.

(L3) $Lxyz \rightarrow (Atx \ \& \ Aty \ \& \ Atz)$.

In words: the linking-relation is confined to atoms.

(L4) $(Lxyz \ \& \ Lxuz) \rightarrow y=u$.

In words: individuals are connected by at most one link.

(L5) $Lxyz \rightarrow (Lxxx \ \& \ Lzzz)$.

In words: all individuals which are linked to something, also link themselves to themselves. We could introduce the terminology 'x is a unit' for the formula '$Lxxx$'. Then (L5) reads: only units are linked to one another.

(L6) $(Lxxx \, \& \, Lyyy) \to (\exists z) Lxzy.$

In words: any two individuals which link themselves to themselves are linked to one another. Or, using the alternative terminology: any two units are linked.

It turns out that these axioms regarding the linking relation render the theory adequate to the model-theoretic interpretation provided for it:

(2.10.4.5) THEOREM: The theory of individual relations is semantically sound and complete.

The main step in the proof of this result is the analog of Lemma (2.5.4.14): If K is a consistent set of formulas such that infinitely many variables are not free in any member of K, then there exists a model M (whose universe is a universe of individual relations) and M satisfies every member of K. Our proof of this lemma is so long and tedious that we hesitate to include it in these pages. In rough outline, we proceed as follows:

As in the proof of (2.5.4.14), extend K to a maximally consistent and omega-complete set K^*. Let A_0 = the set of all singletons of pairs of variables; let F = the set of all unions of non-empty subsets of A_0, and let $\mathbf{R}$ = the inclusion relation restricted to F.

Let $\mathbf{U}$ = the family of all non-empty sets x such that for some variable α, x = the set of all pairs (y, z) such that for some variable β, $\ulcorner \beta \leqslant \alpha \, \& \, Ly\beta z \urcorner \in K^*$. Let $\mathbf{A}$ = that function whose domain is the set of all variables α such that $\ulcorner \alpha \circ \alpha \urcorner$ is in K^* and for every such α, $\mathbf{A}(\alpha)$ = the set of all pairs (y, z) such that for some β, $\ulcorner \beta \leqslant \alpha \, \& \, Ly\beta z \urcorner \in K^*$. It turns out that $\mathbf{M} = \langle \mathbf{R}, \mathbf{U}, \mathbf{A} \rangle$ is a model for K^*. It helps to observe that x is $\mathbf{R}$-least in $\mathbf{U}$ just in case for some y in $\mathbf{U}$ and for some pair (u, v) in y, x = the set of all pairs (a, b) such that $\ulcorner a = u \, \& \, b = v \urcorner \in K^*$.

Although universes of individual relations were so defined that they comprise individuals which are literally relations in the set theoretic sense, it is natural to admit universes of the same structure whose individuals need not be thought of as sets of ordered pairs. Instead of construing our theory to concern sets of ordered pairs, we may think of it as comprising assertions about sums of concrete individuals whose left and right sides are distinguished. Units are regarded as concrete objects in the usual sense; and a link between two units x and y is thought of as a slightly unusual object which shares its left side with x and its right side with y.

The theory of individual relations admits of various definitional extensions. For example, it seems natural to define the concatenate or pair (x, y) of two units x and y as follows:

(2.10.4.6) $(x, y) = z \leftrightarrow (Lxxx \ \& \ Lyyy \ \& \ (\forall u) \ [Atu \rightarrow (u \leqslant z \leftrightarrow [u = x \lor u = y \lor Lxuy])])$.

In words: z is the pair (x, y) just in case both x and y are units, and all and only those atoms are part of z which are either identical with x or y, or link x to y.

Thus, if x and y are units, then the pair (x, y) will comprise, in addition to x and y, just the link between x and y. Also, the pair (x, x), where x is a unit, will comprise x and that atom which links x to x. But the latter, by (L4), is just x itself; so that the pair or concatenate (x, x) is just x itself. Due to (L1), we have, with respect to units x and y, the law:

(2.10.4.7) $(x, y) = (u, v) \rightarrow (x = u \ \& \ y = v)$.

However, with respect to longer sequences of units, we cannot proceed in this manner. For example, the three-term sequence of units (x, y, x) will intuitively have for parts just x, y, and the links between x and y and between y and x. But the triple (y, x, y) has exactly the same atomic parts as (x, y, x) and is therefore identical with it. It seems, therefore, that sequences of three or more units can be defined only under the convention that the terms are pairwise distinct. However, it seems reasonable to regard items which repeat themselves as abstract; and given a nominalist's suspicion of abstract entities, it does not appear counter-intuitive that sequences of repeated units should not be admitted as individuals. With this understanding, the definition can be given as follows:

(2.10.4.7) $(x_1, ..., x_n) = y \leftrightarrow [(Lx_1x_1x_1 \ \& \cdots \& \ Lx_nx_nx_n)$
$\& \ (x_1 \neq x_2 \ \& \cdots \& \ x_1 \neq x_n) \ \& \ (x_2 \neq x_3 \ \& \cdots \& \ x_2 \neq x_n)$
$\& \cdots \& \ x_{n-1} \neq x_n \ \& \ (\forall u)(Atu \rightarrow [u \leqslant y \leftrightarrow (u = x_1 \lor \cdots \lor u = x_n \lor Lx_1ux_2 \lor \cdots \lor Lx_{n-1}ux_n)])]$.

In words: y is the ordered n-tuple $(x_1, ..., x_n)$ just in case each x_i is a unit, all the x's are pairwise distinct, and exactly those atoms are part of y which are either identical with some x_i or else link some x_i with its successor.

In treating of concatenates, we did not find it necessary to assume that all sums of individuals exist. Indeed, this assumption would be undesirable with respect to concrete individuals. However, it seems plausible to stipulate that, whenever a certain string exists then all shorter (non-empty) strings contained in it shall also exist. Hence, for all $1 \leqslant i < j \leqslant n$, it shall be a postulate that

(2.10.4.8) $\quad (\exists y) \, y = (x_1, ..., x_n) \to (\exists y) \, y = (x_i, ..., x_j)$.

Sequential individuals satisfy the expected law that two of them are identical just in case they are identical term by term:

(2.10.4.9) $\quad$ If $(x_1, ..., x_n) = (y_1, ..., y_m)$, then $n = m$ and for all $i \leqslant n$,
$$x_i = y_i.$$

Let $X = (x_1, ..., x_n)$ and $Y = (y_1, ..., y_m)$ and assume that $X = Y$. Call z a *unit* if $Lzzz$, and call z a *link* if for some i, $Lx_i z x_{i+1}$ or $Ly_i z y_{i+1}$. No unit is a link; for, assuming that $Lzzz$ and $Lx_i z x_{i+1}$, it would follow, by (L1), that $x_i = z = x_{i+1}$, contrary to the hypothesis and the definition (and similarly if x's are replaced by y's). Hence, X has for atomic parts exactly n units, and Y has m. Thus, $m = n$. Claim: for each i, $x_i = y_i$. Suppose not, and let i be the least index such that $x_i \neq y_i$. Case A: $i = 1$. For some u, $Lx_1 u x_2$. Since u cannot be a unit, for some j, $Ly_j u y_{j+1}$. By (L1), $x_1 = y_j$ and $x_2 = y_{j+1}$. By hypothesis, $j \neq 1$. Hence, for some v $Ly_{j-1} v y_j$. Therefore, for some k, $Lx_k v x_{k+1}$. By (L1), $x_k = y_{j-1}$ and $x_{k+1} = y_j = x_1$. Hence, $Lx_k v x_1$: impossible. Case B: $i \neq 1$. For some u, $Lx_{i-1} u x_i$. Hence, for some j, $Ly_{j-1} u y_j$. By (L1), $x_i = y_j$ and $x_{i-1} = y_{j-1}$. Since i is the least index of the sort mentioned, $x_{i-1} = y_{i-1}$. Hence, $y_{i-1} = y_{j-1}$, $y_i = y_j$, and $x_i = y_i$.

Sums of individual pairs have much the same properties as do sets of pairs; that is, relations. For example, if x, y, u, and v are distinct units, and so are a, b, c, and d, and if the sum of the pairs (x, y) and (u, v) is the same as the sum of the pairs (a, b) and (c, d), then (x, y) is identical with one of the pairs (a, b) or (c, d), and (u, v) is identical with the other one. It appears that individual pairs can be uniquely reconstructed from a sum of pairs.

Among the four methods, discussed in this chapter, which would permit a nominalist of Goodman's persuasion to admit sequential individuals, the last one seems most advantageous. It is the only one among the four which justifies any hope that relations can be characterized in terms of sequential individuals. For it is the only workable theory among the four which permits us to regard both pairs (x, y) and (y, x) as existing individuals. Unlike the approach briefly discussed in Section 2.10.2, our last method made it possible to characterize sequential individuals without recourse to an auxiliary semantical notion, such as that of 'being immediately to the left of'. Greater simplicity was obtained, in the definition of models, by merely delimiting a suitable class of part-whole relations. True, our intuitions were strained a bit since we had to accept as new individuals ones whose halves were carved out of the flanks of old individuals. But this would seem more palatable than the admission of abstract positional parts.

REFERENCES

[1] Goodman (1966), p. 192.
[2] Goodman (1956). The relevant passages are not too clear, notably the one beginning with the sentence (p. 18): 'Whatever can be construed as a class can indeed be construed as an individual, and yet a class cannot be construed as an individual.' This sentence is incompatible with the assumptions that there is a class and that every class can be construed as a class. Thus, it might seem that Goodman denies the existence of classes. Yet he also holds that individuals can be generated relative to the ancestral of membership; and if individuals, so construed, are not also classes, I would not know what they are.
[3] This, at any rate, is my understanding of the cryptic analogy occurring in Goodman (1956), p. 18.
[4] Quine (1953), p. 114.
[5] For the purpose of stating informal examples, we shift here and elsewhere from the set theory officially endorsed in Section 1.3 to a set theory with individuals.
[6] This difficulty is raised in Hempel (1957).
[7] Goodman (1958).
[8] Goodman (1966).
[9] See, e.g., Halmos (1963), p. 70, Theorem 5. Actually, it is not the part-whole relations themselves which are Boolean algebras, but rather certain structures whose constituents are uniquely determined by a given part-whole relation. These respective constituents are the field of a part-whole relation and the operations assigning suprema, infima (suprema of common parts), and complements (suprema of non-parts).
[10] In subsequent developments it will be found that this interpretation accords with the intuitive identification of the universes of models with 'possible worlds', a move which has proved to be philosophically fertile.
[11] Goodman (1966), p. 53.
[12] Goodman (1956), p. 25; also in (1966), p. 51.
[13] The first system, which may be called a 'calculus of individuals' in the indicated sense, was published by Leśniewski in 'O Podstawack Matematyki', *Przeglad Filozoficzny*, **30–34** (1927–1931) as a deductive system called by him 'Mereology'. However, mereology was intended to have Leśniewski's 'ontology' as a subtheory, thereby strengthening the combined theories. Tarski published a simplified version of Leśniewski's calculus as an appendix to J. H. Woodger's *Axiomatic Method in Biology*, Cambridge 1937, Appendix E, p. 160. Another version has been published by Tarski in *Logic, Semantics, Metamathematics*, Oxford 1956, p. 25. Further investigations into Leśniewski's mereology are due to Sobociński (1955) and Clay (1965), and extensive

discussions and further references can be found in E. C. Luschei, *The Logical Systems of Leśniewski*, Amsterdam 1962.

A version of the calculus developed independently of Leśniewski by Henry S. Leonard and Nelson Goodman was included in Leonard's thesis, *Singular Terms*, typescript, Widener Library, 1930; and a later version was published by these authors in 'The Calculus of Individuals and its Uses', *Journal of Symbolic Logic* 5 (1940) 45–55.

Goodman has later published an outline of a calculus of individuals in *The Structure of Appearance*; 1st edition, Cambridge, Mass., 1951, pp. 42–55; 2nd edition: (1966), pp. 46–61. In this presentation, Goodman claims not to give a complete account of the calculus, but it is not made clear whether he would wish to add further axioms to this later version of the calculus, and whether or not he still endorses in its entirety the earlier version which involved reference to classes. Also, the underlying logic of this later system, notably the intended treatment of non-denoting descriptive phrases, has not been made fully explicit, but is probably akin to Russell's. Thus, either the reflexivity of identity or existential generalization or both fail to hold in this system.

[14] Such a scheme was in effect proposed by Goodman (1966), p. 52.

[15] Terms are mentioned in anticipation of enrichments of the language.

[16] See Eberle (1969b).

[17] See, e.g., R. Sikorski (1964). I owe to Alfred Horn, in oral communication, the verification of the independence by means of such an algebra and in a system differing from LGCI only by the inessential assumption that all variables have values.

[18] A treatment of some atomistic calculi akin to CII, II, and III, can be found in Eberle (1967). The treatment there is quite condensed, the systems have existential implications (that is, their theorems hold only in non-empty universes), and the completeness proofs given in that paper (which owe much to Professor David Kaplan) are more elegant than the ones provided here, but less easily extendable to other systems which will be treated in the present work.

[19] For some of the standard proofs, see, e.g., Mendelson (1964). For modifications, required because we admit empty universes and variables without values, see Eberle (1969b).

[20] For analogous details, see Eberle (1969b).

[21] Although the only terms are variables, we refer here, as earlier, to terms in order to facilitate extensions of the language. Thus, we might extend CI II by defining appropriate operational terms for 'the sum of x and y' and 'the product of x and y'.

[22] At the end of Section 2.2 we have briefly observed that every part-whole relation, upon addition of a null-element to its field, determines a certain Boolean algebra. The models of CI II, upon addition of a null-element to the universe, are Boolean ideals in such algebras. See Halmos (1963), section 11.

[23] Goodman (1966), p. 117.

[24] M. G. Yoes, Jr. (1967). For a discussion ensuing from that paper, see Eberle (1968), Schuldenfrei (1969), and Eberle (1969c).

[25] The assumption of atomicity with respect to the ancestral of membership is made in set theory by adopting the Axiom of Regularity (or, of Restriction): every non-empty set of sets is disjoint from one of its elements. Given the Axiom of Choice, the Axiom of Regularity is equivalent to the assertion that there is no function f which is defined on all numbers and such that $f(n+1)$ is always a member of $f(n)$. In this sense, the axiom implies that there are no infinitely descending membership-chains, and hence that there are no sets which are non-atomistic with respect to the ancestral of membership.

[26] Hempel (1953), p. 110.
[27] Goodman (1956), p. 24.
[28] Goodman and Quine (1947), p. 112f.
[29] Goodman (1956), p. 24.
[30] An interpreted formal system of logic of this sort which is both sound and complete is characterized in Eberle (1969b).
[31] See Goodman (1966); especially chapters VI,4 and VII,2.

CHAPTER 3

ONTOLOGICAL COMMITMENT AND DESIGNATA OF EXPRESSIONS

We have previously observed that an interpreted formal system cannot be regarded as a reconstruction of nominalism if it implies the existence of entities other than individuals. At least two points must be clarified if this requirement is to be clearly understood: (1) an explication must be provided of what is meant by an individual, and (2) criteria must be presented which determine whether a given interpreted formal system does or does not imply the existence of certain kinds of entities. Chapter 2 has been devoted to the first of these two points. We now turn to the second point: criteria are to be examined which allow us to assess the ontological implications of various systems; notably those which have been developed in the previous chapter, and various extensions of these calculi yet to be formulated.

3.1. ONTOLOGICAL IMPLICATIONS OF THEORIES

A well-known criterion for assessing the ontological commitment of a theory is due to Quine. We state and discuss this criterion in the following formulation[1]:

(3.1.1) 'An entity is assumed by a theory if and only if it must be counted among the values of the variables in order that the statements affirmed in the theory be true.'

If a theory is to be assessed by employing this criterion, then at least the following points need to be clarified:

(1) It is not immediately obvious in which of the following senses the word 'theory' (in 3.1.1) is to be taken:

(a) Quine makes some remarks which suggest that a theory is a standard model. In this sense, e.g., the theory of natural numbers may be identified with that model whose universe comprises all natural numbers and in which the appropriate operation symbols '0', '′', '+', and '.' refer, respectively, to the actual number zero and to the intended operations of successor, addition, and multiplication. This interpretation of the word 'theory' is suggested by Quine's repeated reference to *the* range of values of a variable, and to *the* universe of a theory, where use of the word 'the' suggests that there is only one universe per theory. Unfortunately, there are theories whose standard models have not been specified, or where it seems arbitrary or controversial which models shall count as their standard models. For example, it would seem quite arbitrary to designate a standard model for a calculus of individuals.

(b) It seems more plausible to take the word 'theory' to refer to some axiomatic system, together with any model of that system. In this sense, e.g., the theory of natural numbers may be taken to consist of Peano's postulates, together with any first-order model of those postulates. In this sense of the word 'theory', number theory has many models (in fact, models of every infinite cardinality) and its variables can be made to range over items which are not numbers. For example, under a customary definition of 'natural number' in set theory, most unit sets of natural numbers are not themselves natural numbers. Yet the universe of a model of number theory may well comprise unit sets of numbers or, for that matter, even items which are not sets. Since numbers need not be counted among the values of its variables, does this mean that number theory is not committed to numbers after all?

Of course, if number theory happens to be appropriately axiomatized with the aid of a special predicate like 'is a number', there must exist items, in every universe of number theory, which satisfy the formula 'x is a number'. But these items need not be numbers. Perhaps Quine would retort that number theory assumed the existence of numbers in precisely this sense: that in every universe there must exist items, of whatever sort, which satisfy the formula 'x is a number'. But then, any nominalistic project which has for its aim to construe Peano's postulates as being true of concrete things, is doomed from the start. For even if it should succeed it must, by the nature of the project, end up with things which satisfy the formula 'x is a number', rather than the formula 'x is concrete', and therefore the theory will continue to be committed to numbers rather than to concrete things. On the other hand, it would seem that a nominalist could eliminate any commitment by number theory to numbers by the simple expedient of axiomatizing the theory without using the predicate 'is a number'.

(c) The word 'theory' might be intended to refer to a particular axiomatic system, together with that meaning of its symbols which is customary in ordinary or scientific discourse. In this sense, Peano's postulates, set forth without the predicate 'is a number', would still customarily be taken as a theory treating of numbers. Or some theory of particle mechanics may, in this sense, commit to the existence of particles which are not formally identifiable as the items in a standard model, or as any item in any model of the system, but rather as those entities which constitute the subject-matter of which physicists intend to treat by means of that theory. A physicist can regard a theory as treating of particles if the predicate 'is a particle' does not occur in the theory; indeed, even if no formula of the theory comes close to being synonymous with 'is a particle'.

(d) The word 'theory' might refer not just to one axiomatic system, but rather to a whole collection of systems all of which are intended to treat of a given subject matter whose description is given only extrasystematically. In this sense one might speak of a theory of classes which finds expression in any one of a variety of formal systems not equivalent to one another. In order to assess the ontological commitment of a theory of classes, in this sense, one is permitted to consider any one among dozens of theories all of which are somehow felt to treat of classes. For reasons which will appear below, Quine's criterion seems best suited for judging of the commitment carried by theories in this rather loose sense of the word 'theory'.

(2) In what sense of the words 'an entity' are we to inquire whether an entity is assumed by a theory? Perhaps (a) the assumption of *particular* entities is in question, such as the existence of Pegasus, or of the number two. Or, (b), we are to judge whether a theory assumed an entity of a certain *kind*, where the kind may be definable by a formula in the object-language of the theory. Thus, if the predicate 'is a mermaid' were in the vocabulary of a theory, one might wonder whether the theory assumes entities of the specifiable kind 'mermaid'. It could be, (c), that one is to judge whether a theory 'assumes an entity' in the sense of presupposing that there exists an item of a classification which is definable in the metalanguage of the theory. In this sense, a standard number system, in whose vocabulary the predicate 'is a number' does not occur, can still be said

to assume numbers in the sense of implying the existence of entities which in the meta-language can be described as numbers. Finally (d) the entities in question might be of a kind which is not describable either in the object-language or in the meta-language of the theory, but only extra-systematically. For example, the predicate 'is abstract', whose meaning is rather foggy, might not occur either in the object-language or in the meta-language of a formal theory. Yet it could be that the criterion is still intended to inform us, for example, that number theory assumes an abstract entity.

(3) What could be meant by the word 'must' as it occurs in the phrase 'must be counted among the values of the variables'? What is definitely not intended is the following material implication: 'If the statements affirmed in the theory are true, then the entity in question *is* counted among the values of the variables'. This latter statement is true if its antecedent is false, and thus application of the criterion (3.1.1) to false theories would give the result that all entities are assumed by such a theory. Possible interpretations of the word 'must' will be considered later on.

(4) Some explanation would be helpful of what is meant by 'affirmed' in 'statements affirmed in the theory'. Alonzo Church[2] takes it

that a language carries the ontological commitment of every sentence which is analytic in the language, i.e., of every sentence whose truth is a logical consequence of the semantics of the language. In some cases it may be sufficient to say that a language carries the ontological commitment of all its theorems.

Disregarding, for the moment, Quine's scruples regarding analyticity, let us follow Church by translating 'statements affirmed in a theory' by 'statements which are analytic in a theory'. Quine's criterion will then read:

(3.1.2)   An entity is assumed by a theory if and only if it must be counted among the values of the variables in order that the statements which are analytic in the theory be true.

If one supposes, as it is customary, that all analytic statements are true, no entity whatever must be counted among the values of variables in order to make analytic statements true. Hence, under this reading, no entities are assumed by theories. We conjecture that either of the following alternatives is intended: (a) It could be that Quine thinks of some 'absolute' notion of truth, to be contrasted with truth in a theory, and in a sense which makes it possible for statements to be analytic in a theory and yet false. It will be somewhat difficult to characterize such a notion of truth, which is not system-dependent, with regard to the statements of logical and mathematical theories. Or, (b), one might omit reference to truth altogether and reformulate the criterion as follows:

(3.1.3) An entity is assumed by a theory if and only if it must be counted among the values of the variables occurring in those statements which are analytic in (or derivable from) the theory.

At any rate, the expression 'statements affirmed in the theory' is definitely not intended to mean 'statements affirmed by a person using the language of the theory'. This remark is supported by Quine's insistence that the criterion is designed to assess commitments implicit in discourse, not the commitment made by a person who avails himself of such discourse.

(5) Note that Quine's criterion cannot serve to determine whether the existence of an entity is *denied* in a theory; only whether its existence is not assumed in that theory. Thus if a theory should contain the assertion 'there are no unicorns', Quine's criterion, as stated, does not tell us that this theory denies the existence of unicorns.

Some clarification of Quine's criterion may be obtained by considering the reasons, adduced by him, for thinking it plausible. He proceeds by showing that all categories of expressions, other than variables in quantificational contexts, are either eliminable in a suitable system or else capable of being construed as denotationless expressions under a suitable interpretation:

Since defined symbols are eliminable, the ontological import of sentences containing defined symbols need not be considered.

Descriptive phrases (in a suitable system) are eliminable in context and need thus not be considered.

Proper names and terms of operational form can be treated as concealed descriptions and hence eliminated in context.

Note that there are theories of descriptions[3] of such a sort that descriptive phrases are not generally eliminable from sentences; and if a theory should not even contain descriptive phrases, then constants are not eliminable (in any sense) in favor of descriptions in that same system. Quine seems to recommend that we assess the ontological import of

ONTOLOGICAL COMMITMENT 107

sentences in such systems indirectly, by finding translations of these sentences among the sentences of some other comparable system where the expressions under discussion are eliminable. This raises problems regarding the adequacy of translations from one system to another.

The principle which seems to emerge from this line of argument is this: whenever the expressions of a certain syntactical category are eliminable, if not in the given system then in some comparable one, then the ontological import of such expressions will be that of the expressions in favor of which they are eliminable. Quine notes that variables are also eliminable in combinatorial logics, but he prefers to consider the import of variables rather than that of combinators.

Having 'eliminated' as much as he finds convenient, Quine is left with primitive predicates, variables, at least one quantifier, and a minimum of sentential connectives.

Among the remaining types of symbols, predicates can be treated as syncategorematic expressions; that is, they can be so interpreted that they do not denote anything, while their occurrence in sentences is still essential in determining whether the sentences are true or false. This can be supported by providing a semantical interpretation for sentences containing predicates in such a way that no extensions are assigned to the predicates. The following example illustrates how this can be done:

(3.1.4)   EXAMPLE: Let the language L comprise the one variable 'x'; the formulas 'x is a mermaid', 'Mary is a mermaid', and for all formulas ϕ the sentences '$(\exists x)\,\phi$'. All formulas of L are assumed to be formulas of the meta-language as well. A universe of discourse for L is any non-empty set. If D is such a universe, satisfaction conditions are specified for all formulas of L as follows:
(1) y satisfies 'x is a mermaid' in D if and only if y is in D and y is a mermaid;
(2) y satisfies 'Mary is a mermaid' in D if and only if y is in D and Mary is a mermaid;
(3) y satisfies '$(\exists x)\,\phi$' in D if and only if there exists an item z in D such that z satisfies ϕ in D.
A sentence shall be true in D just in case every element of D satisfies the sentence in D.

Languages with infinitely many variables and finitely many predicates of arbitrary degree can also be interpreted by assigning no extensions to predicates, though such an interpretation will be somewhat more complicated. The given example also serves to illustrate how sentences and logical symbols can be interpreted syncategorematically; that is, without employing in the definition of truth an assignment of entities to these expressions.

Quine's second principle appears to be this: expressions which can be treated syncategorematically fail to contribute to the ontological import of sentences in which they occur. So far, we have understood a syncategorematic interpretation of expressions to be this: a specification of truth conditions for all sentences in which the given expressions occur in such a way that no assignment was employed, in stating these conditions, which assigns certain entities to the given expressions. Note, however, that no assignment to variables was employed in the last example. Why then should not variables also be syncategorematically interpretable, and hence fail to carry ontological import? Indeed, in one sense, bound variables are paradigm cases of syncategorematic expressions: they are mere 'dummies' or 'indicators of cross reference' without denoting in the usual sense. Perhaps the main point is that quantification, unlike e.g. predication, has to be construed relative to some universe of discourse, *provided* that all names have been eliminated, or at least enough of them to prevent that every intended item of discourse bears a name. For otherwise, existential statements can be interpreted in terms of substituents on variables, and without reference to a universe, in the following pattern:

(3.1.5)   '$(\exists x)$ x is a mermaid' is true if and only if for some name N, 'N is a mermaid' is true.

Rather than to speculate further on what precisely the reasons may be for thinking (3.1.1) plausible – an inquiry which would seem to require exploration of various methods for interpreting quantificational theory – we shall attempt to formulate a criterion whose indebtedness to Quine remains problematic.

To begin with, we find it helpful to distinguish informally between two related problems: (1) There is the problem of assessing the *ontological implications* of a theory. Roughly speaking, this is the problem of determining which entities or kinds of entities must exist if the statements asserted in a given theory and under a given interpretation are to be true. (2) There is the problem of assessing the *ontological commitment* of a theory. Roughly, this is the problem of determining which entities or kinds of entities must exist if the statements asserted in a given theory are to be true under some reasonable or natural interpretation (not necessarily that provided by the given theory). Very loosely: the ontological commitment implicit in discourse is the reference to entities which is *unavoidable* under all reasonable interpretations of that discourse. Thus, the assertion 'some zoological species are cross-fertile' commits us to recognizing abstract entities "... at least until we devise some way of so paraphrasing the statement as to show that the seeming reference to species was an avoidable manner of speaking".[4] One is not committed to the given interpretation of a theory if some other interpretation is available which seems just as plausible as the given one. On the other hand, we shall say that the assertion 'some zoological species are cross-fertile' implies (rather than commits us to) the existence of species if, under the given interpretation, reference to species is in fact made, and whether or not such reference is avoidable. Quine seems to address himself to the problem regarding the ontological commitment, rather than the ontological implications, of theories.

In the special case where a theory is understood to be a formal system with a formal (e.g., model-theoretic) interpretation, it seems more fruitful to inquire about the ontological implications than about the ontological commitment of such a theory. As long as the interpretation of sentences is not definite, as may be the case when we do not know the exact intentions of a person asserting a sentences, there can be reasonable doubt concerning the existential import of the assertion. And if it is problematic how an assertion made in ordinary or scientific discourse would be translated into a formal system, there can be question of 'construing' the assertion in different manners. In such cases one may charitably assess the assertion by using that permissible interpretation according to which the greatest number of expressions are treated as syncategorematic. However, if a theory is taken to be a rigorously interpreted formal system (such as the systems of which we have treated), where the expressions in every sentence receive their reference through the given precise interpretation, there is no room for 'construing' an assertion in different manners. Also, it appears possible that every assertion whose interpretation is sufficiently indefinite to admit of 'construing' is such that, given enough ingenuity, every expression in that assertion can be construed as syncategorematic. Thus, it appears doubtful whether reference to an entity is ever unavoidable and whether, in this sense, a theory can ever carry an ontological commitment.

As a first step in what we hope to be the right direction, we formulate without precision the following criterion:

(3.1.6)   Suppose that S is an interpreted formal system and that ϕ is a sentence of S. Then ϕ implies in S the existence of an entity of a certain kind if and only if, according to the given interpretation of S, an entity of that kind must exist if ϕ is to be true in S.

No mention is made, in this formulation, of variables or their values. An informal example may illustrate the intended content of (3.1.6). Suppose that we are given an interpreted formal system S and that

(D)   There exists a dog

is a sentence of S. Then (D) implies in S the existence of a dog just in case the interpretation of S is such that (D) would be false if there were no dog.

In this example, the word 'must' of (3.1.6) has been paraphrased by a subjunctive conditional statement. Before an attempt is made to eliminate these objectionable phrases, it should be pointed out in what respects they are intuitively unclear.

Suppose that the meta-language of an interpreted system is such that the existence of entities of certain kinds is provable quite independently from the interpretation given to S. Thus, it may be provable in a set-theoretic meta-theory that an empty set exists, although no reference to such a set is made in interpreting a given system S. In this case, an empty set must exist; hence, an empty set must exist if (D) is to be true. Thus, (D) implies the existence of an empty set. This result is intuitively unacceptable.

Furthermore, even if we exclude entities whose existence is provable in the meta-theory irrespective of the truth or falsity of statements expressed in the object-language, there remain counter-intuitive instances. With reference to the previous example, suppose that it is provable in the meta-language of S that there exists a dog if and only if there exists a unit-set of a dog. Since a dog must exist if (D) is to be true, and since a unit-set of a dog must exist if a dog exists, it appears that a unit-set of a dog must exist if (D) is to be true. By (3.1.6) one would have to conclude that (D) implies the existence of a unit-set of a dog, contrary to one's intuition.

Cartwright[5] has proposed a criterion which seems to us far clearer than Quine's, but concerning which similar difficulties arise. It is the following:

(3.1.7)   'An elementary theory, **T**, presupposes objects of kind **K** if and only if there is in **T** an open sentence ϕ having α as its sole free variable such that (1) $\ulcorner(\exists\alpha)\phi\urcorner$ is a theorem of **T**; (2) it follows from the semantical rules of **T** that for every x, ϕ is true of x only if x is a member of **K**.'

Note, to begin with, that Cartwright's criterion refers to given semantical rules of a theory. For this reason, the notion which he seeks to explicate seems to be closer to our concept of ontological implication than to Quine's notion of ontological commitment.

Cartwright has attempted to capture the intended meaning of Quine's vague expression 'must' by the clearer expression 'it follows from the semantical rules of **T** that'. This rendition makes it obvious that expressions like 'ϕ implies in **T** the existence of an entity of kind **K**' are non-extensional contexts with respect to **K**. Thus, if some sentence, say 'there exists an egg-laying mammal', should imply the existence of an egg-laying mammal, and if the class **K** of egg-laying mammals should be identical with the class **L** of duck-billed mammals, it should not follow (by an interchange of identicals) that the given sentence implies the existence of duck-billed mammals. By Cartwright's criterion, an interchange of the class name '**L**' for '**K**' in the given context is warranted only if the identity '**K**=**L**' is provable on the basis of the semantical rules.

Let **T** be the elementary theory of natural numbers, let **K** be the set of all sets of numbers, let **K**′ be the set of all ordinals less than a given infinite ordinal, and let ϕ be the formula 'y is a number'. In a set-theoretic meta-theory, every number may be defined as a set of numbers;

in fact, as the set of its numerical predecessors. Also, the notion of an ordinal number may be so defined that all natural numbers are ordinals. Given appropriate semantical rules (and a standard universe of discourse), it will follow from the semantical rules of the meta-theory that whenever 'y is a number' is true of an object x in the universe, then x is in **K** and also in **K'**. By Cartwright's criterion, we may infer that the elementary theory of natural numbers presupposes objects of the kind 'sets of numbers' and ones of the kind 'ordinals'. This consequence is a bit odd; but the presystematic sense of the word 'presupposing' is not clear enough to let us decide whether or not it is counter-intuitive.

Suppose next that **T** is a theory whose unique postulate is the atomic sentence:

(F)   Fido is a dog.

In standard systems of quantification, with which Cartwright is concerned, (F) implies:

(D) $(\exists x)$ x is a dog,

and hence **T** will presuppose, by his criterion, an entity of the kind 'dog'. Consider next a theory **T'** which consists of the same postulate (F), interpreted exactly as it was interpreted in **T**, with only this difference: that **T'** contains no variables and no quantificational contexts whatever, so that (D) is not a theorem of **T'**. It seems reasonable to suppose that **T'** has still the same ontological presuppositions as did **T** (since **T**'s postulate was quantifier free). But to determine this, we need a criterion which does not depend, as do Quine's and Cartwright's, on the occurrence of quantified sentences in theories. Instead, a criterion should be formulated which enables us to assess the ontological implications of atomic sentences directly.

In standard systems of logic, the sentence 'Fido is a dog' implies the existence of a dog because all names, in such systems, are interpreted as denoting names. If the semantical rules admit of denotationless names, as they do in some current systems of logic[6], then 'Fido' need not denote anything, the inference from 'Fido is a dog' to 'there exists a dog' will not be valid, and 'Fido is a dog' no longer implies the existence of a dog. Thus, the existential implications of a sentence seem to depend on the tems which are *designated* by certain expressions in the sentence.

Consider next whether not only names but also predicates can have ontological import, provided they enter a relation of designation. In considering this question, we should not be overly impressed by the manner in which persons, in ordinary discourse, intend to imply existential assertions. For all we know, it may be a distortion of ordinary discourse to interpret predicates as designating expressions; and, if so, the meaning of sentences in theories which do treat predicates as designating expressions may not have obvious counterparts in ordinary discourse. Nor, on the other hand, should we be overly impressed by the manner in which predicates are interpreted in fashionable logical systems. As we have just observed, the fact that names may or may not have ontological import will not be noted by considering standard systems of logic; and our outlook may be similarly biased with respect to predicates. Instead, we shall proceed, as we did in connection with names, by imagining a theory so interpreted that some predicates may designate, while others need not. While we have to imagine such a theory for the present, in Chapter 4 a formal characterization of such a theory will actually be given.

In systems which allow for denotationless names, one can employ a special predicate, 'exists', which, when appended to a name, forms a sentence of the object-language whose truth implies that the name is a denoting name. Thus, the sentence 'Fido exists' will imply that the name 'Fido' denotes something in the universe. It should be similarly possible to employ a special expression which, when concatenated with the predicate 'is a dog' implies that the predicate enters a designation relation. For the sake of grammaticalness, the special expression in question should be a name. Thus, we shall need an individual constant, say 'S', of such a sort that sentences of the form 'S is a dog' are true if and only if the predicate 'is a dog' designates some item or other. If the suggestion is not too shocking, the symbolic name 'S' may be translated into English by the word 'something'. Like the special predicate 'exists', the special name 'something' should be different from, though related to, the existential quantifier. Now, just as the sentence 'Fido exists' should imply the existence of whatever 'Fido' designates, our new sentence 'S is a dog' should imply the existence of whatever the predicate 'is a dog' designates. For, under the interpretation we have in mind, it follows from the semantical rules that the sentence 'S is a dog' is true only if the entity exists which is designated by the predicate.

Suppose that, under a given interpretation, the predicates 'is a dog' and 'is a dragon' either fail to designate altogether or else designate their extensions, that is, the set of all dogs and the set of all dragons in the universe. Then it should be possible to employ the sentence 'Fido exists and Fido is a dog' to imply that there exists a dog; and similarly, to imply by the sentence 'S is a dog and Fido is a dog' that there exists a class of dogs. On the other hand, if both the name 'Hydra' and the predicate 'is a dragon' fail to denote altogether (fail even to denote the empty set), then the sentence 'Hydra is a dragon', unlike 'S is a dragon', would not presuppose the existence of a set of dragons.

Now, under classical model-theoretic interpretations of formal systems, all names are interpreted by assigning to them certain items, and all predicates are interpreted by assigning to them their extensions. Hence, under this interpretation, the sentence 'Hydra is a dragon' seems to imply the existence of an item denoted by 'Hydra' and the existence of the set of dragons. If this sentence is not, in ordinary discourse, felt to have this implication, we take this to show that classical model-theoretic interpretations depart to that extent from the interpretations which are natural for ordinary discourse. If Cartwright allows for such standard interpretations of elementary theories, we would have to conclude, contrary to his criterion, that sentences like '$(\exists x)$ x is a dragon' also imply entities of the kind 'sets of dragons'.

Quine himself, by implication, seems to agree with this conclusion. For he points out[7], as a possible objection to his proposal, that there might be an occasion which forces us to speak of extensions of predicates; namely, in defining a semantical notion of logical truth. He meets this threat by mentioning the completeness result which enables us to replace the semantical notion of logical truth by the syntactical notion of provability which does not require an assignment to predicates. He seems to feel then, that *if* there were a need for assigning extensions to predicates, then his opponents would have some justification in taking predicates to have existential import.

We have arrived at the conclusion, roughly, that sentences imply the existence of those items which must be designated, according to a given interpretation, by the expressions in the sentences, if the sentences are to be true. Any criterion which may be formulated along these lines must presuppose the notion of *designation* (just as Cartwright's criterion presupposes the notion of a semantical rule). Of course, a semantical relation may in fact be a designation relation without being so called, while not every semantical relation into which expressions may enter qualifies as a designation relation. Inventiveness in the construction of semantical systems prevents that one defines the desired notion both precisely and with sufficient generality to pick out the intended relations in every conceivable semantical system. Instead, let it suffice to say that designation relations are ones into which expressions enter, ones which are typically employed in defining the notions of truth and of satisfaction in the non-intensional part of a theory, and ones which are in some sense non-derivative and essential if all other semantical notions of the theory are to remain unchanged.

In model-theoretic interpretations akin to those mentioned in Section 1.3.2, the functions which assign to names the things named, the functions which assign to predicates their extensions, and the ones which assign to variables their values, would all be regarded as designation relations during the present discussion. This example shows that we intend the word 'designation' to be taken in a rather wide sense perhaps better expressed by the terminology 'referential assignment'. Intended is the kind of assignment to expressions which, unlike intensional assignments, varies from one universe of discourse or from one model to another.

It is possible to employ model-theoretic methods which avoid explicit use of such functions. For example, Tarski[8] uses relational systems of such a sort that the relations which interpret predicates are themselves constituents of the system, while the connection between a particular predicate and the relation meant to interpret it, is given simply by appropriate cross-indexing. Similarly, assignments to variables are replaced by infinite sequences of objects, where the ith object in the sequence is treated as the value of the ith variable. Although no extension assignments or assignments to variables are here mentioned as such, we would still regard as designation relations the relation between the ith predicate and the ith relation of the system, and that between the ith variable and the ith term in a sequence satisfying a formula.

Before we formulate our criterion, another minor difficulty needs to be mentioned. Cartwright's explicandum has the form 'T presupposes objects of kind **K**'. It turns out that '**K**' refers to classes, say to the class of all dogs. By literal substitution, one obtains explicanda of the form

'T presupposes objects of the kind the class of all dogs', which is not English as it stands and misleading if taken to mean that T presupposes sets of dogs. Instead, T is here intended to presuppose members of the set of dogs; that is, objects of the kind 'dogs'.

We shall presently risk our own formulation:

> (3.1.8) Suppose that S is a formally interpreted system of such a sort that the notion of an admissible universe of discourse for S (or of a model) has been defined, and that for every such universe the notions of truth and of designation (of referential assignments) with respect to the universe have been specified. Let U be an admissible universe of discourse of S, and let ϕ be a sentence of S. Then:
>
> The truth of ϕ in U implies (a) the existence of and (b) the non-existence of an element of K exactly when there is an expression ξ in ϕ such that the following is provable in the meta-theory of S: ϕ is true in U just in case (a) there exists or (b) there does not exist a member x of K such that x is designated by ξ.

A few examples may help. Suppose that some interpreted formal system S, together with the sentence

(H)   Hydra is a dragon

satisfy the hypothesis of (3.1.8) and that U is some admissible universe of discourse. Then,

> The truth of (H) in the universe U implies the existence of an element in the class of all dragons in U exactly when there is an expression in (H) [say, the name 'Hydra'], such that, provably, (H) is true in the universe U just in case that expression [say, 'Hydra'] designates some member in the class of all dragons in U.

Informally, (H) will imply the existence of a dragon just in case (H) is true exactly when it makes reference to a dragon. Next, the sentence

(D) $\sim(\exists x)\, x$ is a dragon

hould deny the existence of dragons. We obtain:

> The truth of (D) in the universe **U** implies the non-existence of an element in the class of all dragons in **U** exactly when there is an expression in (D) [say, the variable 'x'] such that, provably, (D) is true in the universe **U** just in case that expression [say, 'x'] designates no member in the class of all dragons in **U**.

Assuming that (D) is provably true exactly when there is no referential assignment of a dragon to the variable 'x', (D) will deny the existence of dragons. Note, incidentally, that if we fix on some syncategorematic expression in (D), such as the left parenthesis, we shall not thereby be able to show that (D) denies the existence of dragons. For the relevant conditional from right to left, 'if the left parenthesis designates no dragon in **U** then (D) is true in U', will fail to be provable. We do however obtain the result that tautologies containing expressions which provably fail to designate deny the existence of everything. If such consequences are not wanted, restrict (3.1.8) to expressions ξ in ϕ other than ones which provably fail to designate in every universe.

Further, assume that **S** is interpreted by the 'classical models' defined in Section 1.3.2, and consider again the sentence (H):

> The truth of (H) in the universe **U** implies the existence of an element of the class of all non-empty extensions assigned with respect to **U** to the predicate 'is a dragon' exactly when there is an expression in (H) [say, 'is a dragon'] such that, provably, (H) is true in **U** just in case that expression ['is a dragon'] designates a non-empty extension assigned with respect to **U** to the predicate 'is a dragon'.

We shall grant ourselves the slight looseness of identifying universes with models. Since the sentence (H) will be true in a 'classical model' just in case the designatum of 'Hydra' is a member of the extension assigned to 'is a dragon', the truth of (H) will imply the existence of a non-empty class assigned to 'is a dragon'. As we have previously noted, this does not seem to accord with the meaning which (H) has in ordinary discourse, although it is a consequence of interpretations by 'classical models'.

So far, we have attempted to define a relation of existential implication which a sentence bears to an entity only if it actually makes reference to that entity. Thus, reverting to an earlier example, if some sentence should imply the existence of a dog and if it should be provable that there exists a dog if and only if there exists a unit set of a dog, the sentence will not imply, by our criterion, the existence of a unit set of a dog unless some expression in it makes actual reference to such a unit set. Also, even if it should be provable that all dogs are individuals, the truth of that sentence will not imply the existence of an individual, because the sentence 'there is a dog' will not be true *just in case* there is an individual. However, we shall need a somewhat more liberal notion, according to which anything which is implied by the implications of a sentence will itself be an implication of that sentence. We call this more liberal notion that of 'indirect implications', and characterize it as follows:

(3.1.9) Under the conventions of (3.1.8), the truth of ϕ in **U** indirectly implies (a) the existence of and (b) the nonexistence of an element of **L** just in case for some **K**,

(a) the truth of ϕ in **U** implies the existence of an element of **K** and **K** is provably included in **L**, or

(b) the truth of ϕ in **U** implies the non-existence of an element of **K** and **L** is provably included in **K**.

Thus, if the truth of a sentence should imply the existence of a unicorn, and if it is provable that all unicorns are either unicorns or dragons, then the truth of that sentence should indirectly imply the existence of something which is either a unicorn or a dragon. And if the truth of a sentence implies that there are no unicorns, while it is provable that all white unicorns are unicorns, then the truth of that sentence should indirectly imply that there are no white unicorns.

So far, our notion of ontological implication has been relative to universes of discourse. We would like to eliminate this relativization to universes. With slight informality, this can be done as follows:

(3.1.10) Adopt previous conventions regarding the system **S** and the sentence ϕ of **S**. Then:

(a) ϕ implies in **S** the existence of a so-and-so just in case for every admissible universe of discourse **U** of **S**, the truth of ϕ in **U** implies the existence of an item in the class of things which are so-and-so relative to **U**;

(b) ϕ denies in **S** the existence of a so-and-so just in case for every admissible universe of discourse **U** of **S**, the truth of ϕ in **U** implies the non-existence of an item in the class of things which are so-and-so relative to **U**;

(c) ϕ implies in **S** the possible existence of a so-and-so just in case for some admissible universe of discourse **U** of **S**, the truth of ϕ in **U** implies the existence of an item in the class of things which are so-and-so relative to **U**.

Thus, if it should be the case that in every universe of discourse **U** the truth of the sentence 'there is a dragon' implies the existence of a member of the class of all dragons in **U**, then that sentence should imply the existence of a dragon. And if that sentence is so interpreted that its truth, in every universe **U**, implies the existence of an item in the class of non-empty extensions assigned relative to **U** to the predicate 'is a dragon', then that sentence should imply the existence of a non-empty extension of that predicate. Similarly, if in every universe **U** the truth of 'there are no dragons' should imply the non-existence of dragons in **U**, then that sentence denies the existence of dragons.

Suppose that the definite description 'the unique thing which is identical with 2' lacks denotation in every universe of discourse other than the one which consists of the number two alone. Then it may be provable, with regard to that one universe, that the truth of the sentence 'the unique thing which is identical with 2 is even' implies the existence of an item which is even and designated by the given description; and in other universes the truth of that sentence may, if suitably interpreted, imply the existence of an even item not designated by that descriptive phrase. In that event, we should not infer that the given sentence implies the existence of an even designatum of 'the unique thing which is identical with 2', since there are universes where no such item is implied. Instead, by part (c) of the previous definition, we shall conclude that the sentence implies the *possible* existence of such an item.

Again, there might be occasion to speak of the indirect implications of a sentence. Thus, while it seems somewhat unintuitive to say (and would not be a consequence of our criteria) that elementary number theory implies the existence of ordinal numbers, we allow that elementary number theory *indirectly* implies the existence of ordinals. The appropriate notions can be obtained by replacing, in (3.1.10), every occurrence of the word 'implies' by 'indirectly implies', and of 'denies' by 'indirectly denies'.

Assuming that a suitable notion of translation from one system to another is avalaible, one can characterize ontological commitment in terms of ontological implication, as follows:

(3.1.11)   Suppose that for any two systems S and S' and for every sentence ϕ of S it is determined which sentences of S' are translations of ϕ. Then a sentence ϕ of S carries an ontological commitment to a so-and-so just in case the following two conditions are satisfied: (1) ϕ implies in S the existence of a so-and-so, and (2) there is no system S' and no sentence ψ of S' such that ψ is a translation of ϕ and ψ does not imply in S' the existence of a so-and-so.

We rather suspect that sentences carry no ontological commitments at all; or that, if they do, it cannot be demonstrated.

Having clarified various notions of ontological import, let us attempt next to give more precision to the nominalistic demand that systems should not imply the existence of any entities other than individuals. With reference to universes of individuals, as previously defined either by (2.3.11) or by (2.8.6), we try out several increasingly stringent demands:

(3.1.12)   A system S is nominalistic just in case there is no theorem ϕ of S which indirectly implies in S the existence of an entity which is not in some universe of individuals.

Due to the mention of theorems, in this condition, it remains possible that some sentence which is true (though not provable) makes reference to non-individuals. Hence, possible extensions of the system, obtained by adding true postulates, could result in a system which is not nominalistic. This demand does not seem strong enough.

(3.1.13)   A system S is nominalistic just in case there is no sentence ϕ of S which is not logically false and which indirectly implies in S the existence of an entity which is not in some universe of individuals.

According to this stronger demand, no sentence whatever shall indirectly imply the existence of non-individuals, barring only those sentences which are provably false in every universe of discourse. The reasons for this exception are these: if a sentence is contradictory and contains an expression which, like the left parenthesis, provably fails to designate anything, then its truth can be shown to imply, by (3.1.8), the existence of anything whatever. This condition still seems too weak. For it is met if some sentence implies the existence of non-individuals in almost all universes of discourse and fails to do so only, say, in the empty universe.

(3.1.14) A system S is nominalistic just in case there is no sentence ϕ of S which is not logically false and which indirectly implies in S the possible existence of an entity which is not in some universe of individuals.

According to part (c) of (3.1.10), this stronger demand is violated by a sentence of a system if there exists even one universe of discourse such that the truth of the sentence in that universe indirectly implies the existence of a non-individual. However, this demand is still too weak. For it is violated by a sentence of a system only if its truth in some universe U implies the existence of an entity which is in no universe of individuals whatever. The condition is not violated if the implied entity happens to be in some universe of individuals other than U. Since any entity will occur in some universe of individuals or other, this restriction is too permissive.

(3.1.15) A system S is nominalistic just in case there is no sentence ϕ of S which is not logically false and no admissible universe of discourse U such that the truth of ϕ in U indirectly implies the existence of an item x of such a kind that if x is in U then U is not a universe of individuals.

This demand is met if every admissible universe of discourse is a universe of individuals and if no expression is construed as designating any entity not in the universe of discourse. In subsequent developments we shall attempt to meet this demand by interpreting all formal systems by means of models whose universe of discourse is invariably a universe of individuals (as previously defined), and such that all referential assignments which may occur in such models shall assign to expressions no items other than individuals in the given universe.

We are aware that the criteria concerning ontological implications which we have formulated are likely to be highly controversial. Indeed, experience indicates that almost all criteria of this generality turn out to be either vague or mistaken. For this reason, we shall attempt to justify all subsequent developments on grounds which are independent of the topics discussed in the present section, although we shall implicitly heed the demands which we have imposed on nominalistic systems.

In the following sections, some traditional proposals will be examined concerning the interpretation of predicates and of other non-logical symbols which one is tempted to treat as expressions designating items other than individuals. In doing so, we hope to learn how traditional nominalism differs from traditional realism with respect to the interpretation of various types of expressions; and these considerations may afford independent grounds for calling just those systems 'nominalistic' which happen to meet the criteria stated in the present section.

3.2. ONE-PLACE PREDICATES AND THEIR EXTENSIONS

In Section 1.2, brief reference was made to one traditional approach to the problem of universals which clearly raises a semantical issue. This approach is exemplified by the following excerpt from Russell's writings[9]:

> ... there remains a metaphysical question, irresistibly suggested by the linguistic distinction of subject and predicate. The word 'Socrates' is a name, and most people think there is no difficulty as to what the word means: it means a certain person who lived a long time ago, and annoyed the Athenians by being too argumentative.... Thus it is clear that the word 'Socrates' means something which is not a word, but a man. But how about 'human'? This is a word, not an empty noise; it certainly has meaning. But is there something that it means?... The metaphysical problem, is to find something which can be meant by words that are predicates – or, at least, this is the form in which the problem first presents itself.

For the purposes of informal discussion, we shall use the term *designation* to refer to that relation between a predicate and an entity which is the approximate counterpart of the relation obtaining between a name and the thing named. Disregarding the fact that Russell uses the word 'meaning' ambiguously to refer to both sense and denotation, his point seems to be the following: one way of introducing the problem of universals is that of inquiring whether predicates designate in much the same fashion in which names denote, and what sort of entities might be designated by predicates.

Realists, approaching the issue of universals in this manner, have generally held that each meaningful predicate, just like every univocally denoting name, designates some unique entity. However, the items designated by predicates are not, in general, members of the same universe of discourse from which things denoted by proper names are drawn. For example, if the names of a given vocabulary are taken to name members of a given universe U, then predicates might designate classes or properties of things in U, where these classes or properties are typically not themselves members of U.

As an example of the manner in which realists interpret predicates, let us return to the notion of a 'classical model' which was defined in (1.3.2.12). Such a model, we recall, was a certain quadruple of the form $\langle U, O, P, A \rangle$, where U is the universe of discourse, O interprets (among other things) names, and P interprets predicates. Simplifying the account a bit, if **n** is any name in the vocabulary of the given language, O assigns to **n** some element of U. In this sense, names are construed as denoting, by means of the function O, some unique entity in U. On the other hand, if p is some one-place predicate belonging to the same vocabulary, then P assigns to p some subset of the universe U (actually the designata are more complex, but this complication may be disregarded in the case of simple names and one-place predicates). Since it is not possible that all subsets of U are again members of U, the designatum, according to the assignment P, of the predicate p may not be in the same universe of discourse relative to which names are construed. Also conversely, the item named by the name **n** in U may not, at the same time, be a subset of U, so that it does not qualify as a designatum of some predicate. In this fashion, both names and predicates are conceived as designating uniquely (both O and P are functions), but the designata of names and those of predicates belong, in general, to different categories. For this reason, it is not permissible to interchange the designata of a name and of a predicate without far-reaching revisions in the logical fabric of a formal language.

Unlike realists, nominalists have tended to favor one of the following alternatives:

(1) The view that predicates fail to designate altogether. On this view, emphatically advocated by Quine, predicates are syncategorematic expressions whose meaning is elucidated by interpreting the contexts (sentences and formulas) in which they occur, but not by means of a designation relation which is defined on predicates.

(2) The position that predicates indeed 'designate', in a manner of speaking, but that the semantical relation in question is structurally different from that obtaining between a name and the thing named and is, furthermore, always a relation into the same universe of discourse from which all names take their values. For example, predicates may enter relations of multiple designation: the relation which serves to interpret predicates is not a (single-valued) function, but it is a relation into the same universe of discourse whose members are denoted by names. For instance, the predicate 'white', on this view, designates each particular white thing. Relations of multiple designation have been systematically investigated by Martin.[10]

(3) The tenet that predicates enter a relation of designation which is exactly like the naming relation, even to the extent that the designata of predicates are the same sort of entities which are also named by proper names. Thus, like realists, nominalists of this persuasion propose to interpret predicates by means of (single-valued) functions; but unlike realists, these designation functions are taken to assign to predicates members of the same universe of discourse into which naming relations are construed. For example, it may be held that the predicate 'white' designates (a) the individual sum of all white things (in the sense of 'sum' clarified in the previous chapter), or (b) some concrete object displayed in some bureau of standards under normed light conditions; an exemplar standardizing the property 'white' much like the standard meter represents the length 'one meter', or perhaps (c) some individual 'quale': the phenomenally given color 'white'. Thus, nominalists share the view that either predicates fail to designate altogether, or else they designate the same individuals which one is also prepared to name by names belonging to the same vocabulary.

The philosophically interesting problem, as we conceive it, is not whether predicates (really) designate, or whether they (really) designate entities of one sort or another. The meaningful question is rather this: whether an entire semantical theory can be constructed which adequately treats at least of all referentially transparent contexts, *if* predicates are construed in the various manners suggested by realists and nominalists. In particular, the proposed treatments of predicates should give rise to a reasonable theory of truth for all sentences of subject-predicate form, they should somehow be extendable from predicates to relation expressions, and they should serve in characterizing the semantical notions of validity and logical truth. In this sense, the various positions of realism and nominalism are here taken not so much as doctrines, but rather as proposals whose systematic import in semantics is open to comparative appraisals.

In this spirit, we shall examine next how various relations of designation can serve in definitions of truth, and whether there might be advantages in favoring one of the realistic or one of the nominalistic proposals.

3.3. TRUTH CONDITIONS AND RELATIONS OF PREDICATION

In the preceding section, we have distinguished various proposals concerning the interpretation of predicates. Among them, let us first investigate that nominalistic position according to which predicates should be construed as syncategorematic expressions. We still limit our discussion to one-place predicates.

Consider how the predicate 'true' can be defined for sentences of subject-predicate form. If the number of such sentences in a given language is finite, and if there exist translations, in the meta-language, of all sentences in the object-language, then the definition of truth can take the form of a conjunction listing all instances of Tarski's material criterion of adequacy.[11] For a language L comprising just one sentence of German, the definition might take the form:

(3.3.1)   'Hans ist bleich' is a true sentence of L if and only if John is pale.

No relation of designation is needed or used in stating this truth condition.

However, if a language should comprise infinitely many predicates which are permitted to have distinct extensions, then Tarski's criterion has infinitely many atomic instances, and it is no longer possible to define the predicate 'true' by conjoining, in a definition of finite length, all of these instances. Also, if the meta-language should be insufficiently stocked with translations of sentences or of predicates of the object-language, then the approach exemplified by (3.3.1) can no longer be followed. Instead, one introduces as an auxiliary notion some relation of designation which is defined on predicates. For example, if predicates are taken to designate classes of objects, if we consider just one name but infinitely many predicates of German, and if the English meta-language contains the quotation name of the compound German expression 'Hans ist', then such a truth condition might read:

(3.3.2)   If P is a predicate of German, then the concatenate of the expressions 'Hans ist' and P is a true sentence of German if and only if John is a member of the designatum of P.

Supporters of that nominalistic position according to which predicates should be treated as syncategorematic expressions cannot avail themselves of designation relations and of truth conditions such as those exemplified by (3.3.2). They face difficulties, therefore, in defining truth adequately for languages with infinitely many predicates or with an insufficient supply of translations in the meta-language. Also, even in the case of languages comprising only finitely many predicates, the truth conditions exemplified by (3.3.2) are advantageous, since they permit a reduction in the number of undefined expressions needed in formulating a theory of truth.

Furthermore, as Quine has pointed out[12], there is at least one other occasion which seems to require mention of designata of predicates. It arises when an adequate notion of validity or logical truth is to be provided. For, a sentence containing predicates is logically true only if it is true under every assignment of appropriate entities to the predicates occurring in it. Quine counters by appealing to the well-known fact that the first-order predicate calculus is semantically complete and argues that therefore, for all practical purposes, the semantical notion of logical truth can be replaced by the syntactical notion of a theorem. This retort does not seem quite satisfactory: not only are there incomplete theories where these two notions fail to coincide and some of which may be of interest to a nominalist, but the semantical notion of logical truth seems important even with respect to complete theories, if only because it enables us to express the result to which Quine appeals.

For these reasons, we are sceptical of the claim that an adequate semantical theory can be developed by treating predicates as syncategorematic expressions, and we shall not further be concerned with that nominalistic position whose advocates urge that we do so.

Let us turn next to the interpretations of predicates which are typically proposed by realists and also to ones which are employed by those nominalists who favor relations of multiple designation. Consider all German sentences of subject-predicate form and implicitly regard the copula 'ist' as an initial part of each predicate. If both names and predicates are treated as designating expressions, and if predicates are construed as designating certain classes (in fact, the classes of all those things to which the predicates are intended to apply), then a realist would typically specify a truth condition of the following form:

(3.3.3)   If N is a name and P is a predicate, then the concatenate of N and P is a true sentence of German if and only if the designatum of N is a member of the designatum of P.

Use of the relation of membership in this condition implies that a sentence of subject-predicate form is true only if the predicate designates some non-empty class. Since we shall have occasion to speak of designata of predicates which are not classes, it seems desirable to generalize from this condition by replacing mention of the membership-relation in (3.3.3) by reference to an arbitrary relation R satisfying the condition that

(3.3.4)   a sentence of subject-predicate form is true if and only if the designatum of the subject-term bears R to the designatum (or, to the designata) of the predicate.

We call such a relation a *relation of predication*. It should be noted that relations of predication do not connect expressions, but rather items which are designated by expressions; such relations obtain typically among extra-linguistic objects.

A nominalist who assigns to a predicate, instead of a class, each item in the intended class, would be inclined to state a truth condition of the following pattern:

(3.3.5)   If N is a name and P is a predicate, then the concatenate of N and P is a true sentence of German if and only if the designatum of N is identical with one of the designata of P.

According to this condition then, predicates occurring in true sentences have usually many designata, they designate the same sort of entities which are also named by names, and the relation of predication is that of identity. In as much as membership and identity are logical relations, the relations of predication employed either in (3.3.3) or in (3.3.5) are logical rather than empirical ones. This is one of the reasons why the realistic and the nominalistic positions to which they are congenial are conveniently discussed at the same time.

We have previously observed that realists tend to interpret predicates in such a fashion that their designata are not, in general, members of the same universe of discourse whose elements are also named by the names of the same vocabulary. Against this realism, a nominalist would urge greater parsimony with respect to the categories of entities designated by expressions. If one decides to let names denote entities of a certain sort (say, physical objects, or classes), then predicates should not designate entities of a different sort (such as classes of physical objects, or classes of classes). Once one has fixed on a certain kind of entities which are regarded as nameable for the purposes of a given discourse by the names occurring in the language of that discourse, then one should not have to draw from yet another category of entities in order to supply predicates with designata. Indeed, if nominalists should succeed in offering an adequate theory of reference which is more parsimonious than the realist's with respect to the distinct categories of entities to which a language makes reference, then such economy would appear preferable.

Furthermore, both the realistic truth condition (3.3.3) and the nominalistic one (3.3.5) in terms of multiple designation have a shortcoming which I find difficult to describe with precision, but which may account for the disappointment which many students express upon their first acquaintance with Tarski's theory of truth. Roughly speaking, it is this: that nothing finds expression in these definitions themselves which would give any indication of how one would go about *finding out* whether a sentence is true or how, if a sentence is true, one might *justify* the claim that it is true. Suppose we were to ask under what conditions the sentence 'John is pale' is a true one. The realistic truth condition exemplified by (3.3.3) indeed 'informs' us that the given sentence is true provided the item designated by 'John' is a member of the class designated by 'pale'. But the definition fails to tell us by what principles, if any, the class of pale things is circumscribed, or what there might be in or about John, mention of which could justify the assertion that he is in the class of pale things. Similarly, an analog of the truth condition (3.3.5) tells us that the given sentence is true if John happens to be one of the items designated by the predicate 'pale'; but it is silent on the point of how the claim might be warranted that 'pale' designates, among other things, the particular person John.

It might be thought, at this point, that the widespread disappointment of novices with regard to such truth conditions stems from unreasonable expectations of what such definitions are meant to express. For example, it might be that a truth condition for atomic sentences is expected to be a description of experimental conditions under which the sentence could be empirically verified. This expectation, of course, should not be carried to a definition of truth. For a sentence, like 'John is pale', ought to be meaningful, and there should be conditions under which it is true, even in possible states of affairs which differ from the actual one with respect to basic natural laws. The conditions under which 'John is pale' is true cannot always be conditions under which the sentence is empirically verifiable, if we have in mind experimental conditions which serve that purpose only due to special physical features of the actual world.

It is also inappropriate to expect of a definition of truth that it should make explicit the meaning or connotation of a sentence, or of its semantically relevant parts. It is generally agreed that a condition determining the truth value or denotation of a sentence need not also determine the sense or connotation of that sentence. Furthermore, even if truth conditions were specified in a manner which makes explicit the meaning of all sentences under consideration, such truth conditions might still fail to indicate how one finds out whether a sentence is true, or how one justifies the claim that it is true. Suppose, e.g., that one identifies the sense expressed by a sentence with a rule (or function) which assigns to every possible state of affairs the truth value 'T' or 'F' which the sentence is to have in that state of affairs.[13] It may then be said that the sentence 'John is pale' is true in a given state of affairs just in case the sense expressed by that sentence assigns to the state of affairs the truth value 'T'. But such a truth condition remains 'uninformative', in the sense we have in mind, unless the sense expressed by the given sentence is so specified that it enables us, in principle, to find out whether the truth value assigned to the actual state of affairs is 'T' or 'F'. For instance, the sense expressed by the sentence 'John is pale' might be specified as follows: it is to be that function which assigns to a given state of affairs the value 'T' if John is in the class of all things which are pale in that state of affairs, and assigns to it 'F' otherwise. We have, in a way, said what the sense of the given sentence is; but the manner in which it was specified is no more informative than the realistic truth condition mentioned earlier.

While it is easy to empathize with philosophers who are disappointed in current semantical definitions of truth, it is quite difficult to find any good reasons for that disappointment. What features should a theory of truth exhibit in order to be deemed an informative theory? At a guess, the feature is this: that the formulation of truth conditions should make explicit some effective rule which enables one to decide whether any given sentence is true upon having initially inspected a manageable finite list of items mentioned by the rule. The customary definitions of truth, with respect to logically compound sentences, do express an effective rule of this sort: the recursive clauses in the definition of truth make it possible to compute the truth of compound sentences, given the truth values of their relevant parts. Assuming that the treatment of atomic sentences were informative, such theories of truth would surely leave nothing to be desired in their application to logical compounds. In appraising semantical definitions of truth, traditionally oriented philosophers have tended to consider especially the manner in which simple atomic sentences are treated; and truisms of the sort '"Snow is white" is true if and only if snow is white' are admittedly not very exciting. By contrast, logicians have been primarily interested in the application of a truth definition to logically compound sentences. But for the purposes of logic it is rather immaterial how those logically uncompound sentences are interpreted which have chiefly attracted the attention of traditional philosophers. Thus, for the purpose of interpreting the lower predicate calculus it would seem to suffice if the truth conditions for atomic sentences were specified, in the style of (3.3.4), by employing an arbitrary relation R of predication instead of the customary relation of membership.

Since it does not seem advantageous to construe predicates as syncategorematic expressions, and since the truth conditions favored by realists and advocates of multiple designation fail to specify effective rules which might aid in determining whether atomic sentences are true, let us turn to the last of the nominalistic positions mentioned in the preceding section and investigate whether its proposals might be better. It will be recalled that nominalists of that persuasion propose that each predicate (e.g., the predicate 'white') shall designate just one individual which might be an individual sum (the sum of all white things), an individual 'quale' (the phenomenal quality 'white'), or some concrete exemplar (the standard white thing).

If it is decided that a predicate like 'white' shall designate the individual sum of all white things, then the relation of predication appropriate to this decision is the relation between parts and wholes. More specifically, among various part-whole relations the appropriate relation of predication will be that one with respect to which the item denoted by 'white' is a supremum of all white things. Thus, the intended truth conditions may have the following instance:

(3.3.6)   'The Taj Mahal is white' is a true sentence if and only if the individual denoted by 'the Taj Mahal' is part of the individual sum designated by 'is white'.

Such truth conditions are plausible, if at all, only with respect to those predicates which, assuming that they apply to a whole, will apply to all parts of that whole, and which, assuming that they apply to certain parts, also apply to a whole of those parts. In other words, if predicates are construed in this way, they cannot serve to pick out the extensions of those properties which are not preserved under division or under composition (relative to the given part-whole relation). Thus, a condition analogous to (3.3.6) would be counter-intuitive for a predicate like 'is square', both because not all parts of a square are square and because not all sums of squares are square.[14] One cannot find a sum of squares which might be denoted by the predicate 'is square' of such a sort that squares, and nothing but squares, are part of that sum. In saying this, we assume that the relation between parts and wholes among geometric figures is the obvious one. At the expense of artificiality, a part-whole relation might be so contrived that all parts of squares are squares, and all wholes of squares are squares. But then that same part-whole relation could not serve as a relation of predication appropriate to other predicates; and hence there would not be one part-whole relation which serves, simultaneously, as a relation of predication in all truth-conditions for atomic sentences. But if several relations of predication are admitted, appropriate to different intended meanings of predicates, then it is no longer possible to decide, just on the basis of the structural description of a sentence, which relation of predication shall serve in defining its truth. Thus, the proposal that predicates shall designate sums of individuals does not seem to give rise to a promising theory of truth.

Other nominalists would have it that the predicate 'white' designates, instead of the sum of all white things, some quality which is given in immediate experience: the quale 'white'. From this point of view, phenomenal qualities of individuals are themselves individuals; and any quality which is fit to be ascribed to a predicate (at least, to a primitive predicate) is a phenomenal individual. If the quale 'white' and white objects are to belong to the same universe of discourse, it is natural to think of objects as the sort of items which are composed of qualia, and to select as a relation of predication that relation which obtains between phenomenal ingredients and wholes composed of such ingredients. An example of a truth condition congenial to this view is the following:

(3.3.7) 'The Taj Mahal is white' is a true sentence if and only if the individual quale designated by 'is white' is part of the individual denoted by 'the Taj Mahal'.

The part-whole relation, in this condition, is intended to hold between phenomenal qualities and sums or 'bundles' of such qualities, each individual being construed as the 'bundle' of those basic qualities which can be truly ascribed to it.

We shall investigate the approach just mentioned in considerable detail later on. For the present, it may suffice to note that formally the theory which ascribed to predicates certain sums (say, the sum of all white things) and the theory which assigns to them certain qualia (say, the quale 'white') boil down to truth conditions which are mirror images of one another. For, upon removal of extraneous aids to intuition, the truth conditions appropriate to the two theories have the respective forms:

(3.3.8) Sentences of subject-predicate form are true just in case the designatum of the subject is part of the designatum of the predicate;

(3.3.9)   Sentences of subject-predicate form are true just in case the designatum of the predicate is part of the designatum of the subject.

And these truth conditions differ only in that the relation of predication is either a part-whole relation or the converse of a part-whole relation. We note also that in both cases the designata of predicates are conceived as individuals which are, in different senses, *abstract*. Thus, if the predicate 'white' is taken to designate the sum of all concrete white things throughout space and time, this composite of concrete things would not itself be termed 'concrete' in customary parlance. Rather, it seems 'abstract' in a similar sense in which the class of all white things is abstract: it is distinguished from other composites by reference to a property common to its ingredients. Or, if 'white' is taken to designate some quale 'white', that quale would be regarded as abstract in a different sense: it is an ingredient common to many individuals. And, while abstract individuals may be compatible with nominalism, it would seem preferable to get along without them.

If predicates are construed as designating qualia, then truth conditions, such as those exemplified by (3.3.7), would seem to be somewhat more informative than the ones considered earlier. For, in order to decide whether the sentence 'The Taj Mahal is white' is true, one would have to inspect, according to that truth condition, just two individuals in the given universe of discourse: the quale 'white' and the Taj Mahal. However, one of the items to be inspected is not a concrete one.

Let us consider, finally, the last of the nominalistic positions mentioned in the preceding section. Recall that nominalists holding that position feel that every predicate should designate a unique concrete exemplar of the quality expressed by the predicate. For example, the predicate 'white' might designate some object of uniform white color displayed in a bureau of standards under normed light conditions. The problem will then be that of specifying a suitable relation R of predication which renders plausible all truth conditions like the following:

(3.3.10)   'The Taj Mahal is white' is a true sentence if and only if the individual denoted by 'the Taj Mahal' bears R to the exemplar designated by 'is white'.

Relations R which satisfy this condition appear to be some relations of resembling or, perhaps, of resembling with respect to color. For, if the predicate 'is white' denotes some concrete exemplar, it seems fitting to specify that all and only those objects are properly called 'white' which resemble, in color, that exemplar.

Nominalism is frequently identified with a certain view regarding the meaning or justification of classificatory statements; namely, that classifications are warranted only by virtue of resemblance relations which all individuals of a given (natural) classification bear to one another.[15] Thus, white things fall under one classification because there is a certain resemblance relation which white things, and none other, bear to each other. If this doctrine is to be relevant to the problem of formulating truth conditions, then it may be identified with the proposal that the relation of predication shall be some resemblance relation. In addition, some connection is needed between the predicate 'white' on the one hand and the resembling white things on the other hand. The connection might be that the predicate 'white' designates the quality class comprising all resembling white things. But then, by our criterion, the position turns out to be a realistic one. Again, the connection of the predicate with reality might be this: that the predicate 'white' multiply designates each thing in the resemblance class. But in that case, the relation of predication will be that of identity (recall 3.3.5) rather than that of resemblance. Thus, if the view under consideration is a nominalistic one and if resemblance is meant to serve as a relation of predication, then the link between the predicate 'white' and reality should be so conceived that the predicate designates a unique white thing which concretely exemplifies the color 'white'. It seems, therefore, that this nominalistic theory of classification, if relevant to the theory of truth, is just the one which would endorse truth conditions of roughly the following sort:

(3.3.11)   'The Taj Mahal is white' is a true sentence if and only if the Taj Mahal resembles (in color) the exemplar designated by 'is white'.

The previously proposed relations of predication, namely membership, identity, or the part-whole relation, have the advantage of being relations of predication whose formal characteristics have been precisely expressed. By contrast, the necessary truths peculiar to the resemblance relation have received very little formal treatment. Hence, if one were to employ such a relation of predication in the framework of a semantical theory, then no proofs could be given, in that theory, which require an appeal to the necessary statements concerning the resemblance relation. In order to make up for this lack, we shall subsequently construct an interpreted formal system treating of the resemblance relation. The axioms of that system may then be incorporated into the meta-theory, and subsequent appeals to these axioms should render meta-proofs formally acceptable.

While an axiomatic treatment of the resemblance relation can be just as rigorous as that given to the membership relation, resemblance has the advantage over membership, as a relation of predication, that it lends itself more easily to an empirical interpretation, and that the items which enter the relation can be conceived as concrete individuals. Furthermore, truth conditions like (3.3.11) would seem to be far more informative that the alternatives previously considered. For the condition expressed there would enable us to find out whether a sentence of subject-predicate form is true by inspecting and comparing just two concrete items: the thing named, and the exemplar designated by the predicate.

Although this last proposal seems very attractive, it leads to serious difficulties which we shall further discuss in the chapter devoted to concreta.

3.4. EXPRESSIONS OF RELATIONS AND OPERATIONS

In Section 3.2, a classification was attempted of various realistic and nominalistic positions regarding the referential status of predicates. The differences between these views arise again from the manner in which realists and nominalists typically propose to interpret relation expressions. Realists are likely to maintain that each relation expression designates exactly one item, namely a relation, which need not be an individual in the sense of occurring in the same universe of discourse to which all those individuals belong which enter the relation. Nominalists, on the other hand, can be expected to hold either that relation expressions fail to designate altogether, or else that they designate individuals.

A similar difference of views must emerge with respect to the interpretation of expressions which are alleged to refer to operations; such as the expressions 'the father of...', 'the intersection of... and ---', and countless others. Although operation symbols (other than constants) are rarely discussed in connection with universals, realists will hold that every operation expression designates a certain item, to wit, an operation on individuals, which need not itself be an individual. And nominalists are presumed to hold that operation expressions, if they designate at all, must designate individuals.

In order to illustrate the realistic approach to the interpretation of relation- and operation symbols, we remind again of the 'classical models' defined in (1.3.2.12). Given such a model, of the form $\langle \mathbf{U}, \mathbf{O}, \mathbf{P}, \mathbf{A} \rangle$, the constituent $\mathbf{P}$ of the model assigns to every **n**-place predicate some **n**-adic relation in the universe **U** of the model; and the constituent **O** of the model assigns to every **n**-place operation symbol some **n**-place operation defined on the universe **U**. Intuitively, a relation expression like 'is warmer than' might be interpreted, in such a model, by having the function **P** assign to 'is warmer than' the set of all pairs $\langle x, y \rangle$, where x and y are in the universe **U** and x is warmer than y. And the one-place operation expression 'the father of' might be interpreted by having the function **O** assign to 'the father of' the set of all pairs $\langle x, y \rangle$, where x and y are in the universe **U** and y is the father of x. It is understood that the sets of pairs designated, according to the assignments **P** and **O**, need not themselves be members of the universe **U** of the model. In this sense, relation- and operation expressions are construed as designating items which are not, in general, individuals.

Without repeating the remarks made in the preceding section, it is clear that the objections urged by nominalists against realists will apply equally to the method which realists use in interpreting relation- and operation expressions. Also, those nominalists who propose to treat predicates as syncategorematic expressions will face similar difficulties in the present context. However, additional difficulties arise for those nominalists who feel that predicates enter relations of multiple designation or that they designate exemplars.

There are no immediate conceptual difficulties in the proposal that a predicate like 'white' shall bear a relation of multiple designation to each white individual. But it is not obvious how this recommendation might be extended to a relation expression like 'is taller than'. If the given expression is meant to designate each individual which is taller than some other, then we are interpreting, in effect, the predicate 'is taller than something' instead of the expression 'is taller than'. This will serve in formulating truth conditions for certain sentences containing the words 'taller than', like 'Tom is taller than someone', but not for others, like 'Tom is taller than Mary'. For similar obvious reasons, it won't do to let the expression 'is taller than' designate instead each thing than which something or other is taller, or each thing which is either taller or less tall than something else. It seems somewhat more plausible to suggest that 'is taller than' shall designate each pair $\langle x, y \rangle$ of individuals x and y, where x is taller than y. However, since nominalists will not designate any items other than individuals, this latter version of multiple designation departs from nominalism, unless ordered pairs of individuals are again construed as individuals. This raises immediately the difficulties regarding ordered or sequential individuals which were discussed in Section 2.10.

Again, there is initial plausibility to the other nominalistic proposal that a predicate like 'white' shall designate a phenomenally given quale 'white', or perhaps a concrete exemplar of the color 'white'. However, the positing of relational analogs of qualia (call them 'relia') taxes the imagination. While individuals can be conceived as bundles of qualia (as sums which have qualia for parts), it is at least odd to conceive of them as bundles which have 'relia' for parts. Or, if we think of the alternative proposal, it seems quite acceptable that there should be some concrete thing, white all over, which standardizes the intended meaning of the predicate 'white'. But the suggestion that there might be a concrete exemplar, displayed in the bureau of standards for purposes of comparison, and perpetually illustrating the correct application of the expression 'x loves y', strikes one as odd, if not obscene.

Although these considerations suggest that nominalists must face additional difficulties in interpreting relation expressions, this should not be taken to show that the difficulties are insurmountable. To the contrary, there are several methods, available to a nominalist, for interpreting relation expressions. We shall presently discuss, in an informal way, a few alternatives which may appear plausible; and in subsequent chapters we shall give formal treatment to one of them.

(1) One might treat relation expressions syncategorematically. This approach leads to difficulties, mentioned in connection with predicates, when languages with infinitely many relation expressions are contemplat-

ed, or if a semantical notion of validity is to be defined. We shall not further consider this approach.

(2) One might admit sequential individuals and have each relation expression designate either a certain sum of ordered individuals or else, severally, the ordered individuals themselves.

In Section 2.10, four possible methods were considered which might enable a nominalist to endorse as individuals certain items which have the essential properties of sequences. Three among those four approaches seemed either to resist consistent formal treatments, or else they implied that of two ordered individuals (x, y) and (y, x), at most one can be actualized in a given universe. But this consequence makes them unsuitable for the purpose of interpreting relation expressions in the manner now under consideration. For, if it should happen that Tom is taller than Mary, then the sequential individual (Tom, Mary) will have to be a member of the given universe in order to serve as a designatum of the expression 'is taller than'; and if, in addition, Mary should love Tom, then the ordered individual (Mary, Tom) will also have to be actualized in order to be designated by the expression 'loves'. Thus, only the last of the four methods for constructing sequential individuals, the one described in Section 2.10.4, can serve nominalists in interpreting relation expressions.

The extra-systematic interpretation which was initially carried to the formal developments in Section 2.10.4 amounted to this: One supposes that a left and a right side can be distinguished of each individual atom, although the two sides of an atom are not in turn individuals. Given any two atoms (x, x) and (y, y) of the normal sort, called 'units', there will exist an additional slightly artificial atom (x, y), called a 'link', which one thinks of as being composed of the right side of (x, x) and the left side of (y, y). For example, one might think of (x, x) and (y, y) as minimally distinguishable color-stretches which are adjacent in a color spectrum, and of (x, y) as the stretch between them. The ordered pair of the respective units (x, x) and (y, y) was then construed as that sum of individuals whose ultimate constituents are just the atoms (x, x), (y, y) and (x, y).

While this conception seems rather natural with respect to concrete individuals which are lined up side by side, it does not appear to be especially pleasing if one thinks of individuals which are spatially or temporally far apart. Thus, an individual link whose left side is part of Plato and whose right side is part of President Nixon makes for an unattractive (and apparently abstract) combination. Yet individuals of just that sort must be admitted if the pair (Plato, Nixon) is to serve as a designatum of the expression 'is wiser than'.

We shall return to this approach in a subsequent chapter devoted to the discussion of concreta.

(3) One can, as before, admit sequential individuals, but employ them in a different way: According to the approach mentioned in (2), one interprets the predicate 'is taller than' by assigning to it either all ordered individuals (x, y) where x is taller than y, or else the sum of all these ordered individuals. Instead, it seems that one could assign to the given expression just one of those ordered individuals which is thought of as exemplifying the relation.

If this approach is workable, it need not be supposed that for every two individuals x and y both of the ordered individuals (x, y) and (y, x) are actualized in the same universe. Instead, we need suppose only that for every relation there exists at least one ordered individual exemplifying that relation. Assuming that the pair (A, B) exemplifies the relation 'taller than', one might then generally specify (by using a relation of relative comparison yet to be analysed) that an arbitrary individual x shall be taller than any other individual y just in case, with respect to size, x compares with y in a similar fashion in which A compares to B. If one need not suppose that both the sequences (x, y) and (y, x) are ever simultaneously actualized as individuals, then one of the other methods (mentioned in Section 2.10) for constructing sequential individuals might serve as well as that of Section 2.10.4.

Furthermore, sequential individuals which serve as exemplars might be so selected that they are close together; so that the scruples need not arise, as they did in the previous approach, concerning links which consist of halves taken from widely separated individuals.

It is clear that certain relation expressions cannot be interpreted in this way; for there are relations of such a sort that whether or not they obtain between two individuals cannot be determined by comparisons or by operational procedures. However, relation expressions whose intended content resists the present treatment might be characterizable in terms of other expressions whose meaning can be determined by comparison with exemplars.

We shall return to the present proposal after having clarified judgements of resembling comparisons which seem to be needed in formulating truth conditions of the sort briefly illustrated above.

(4) One might refuse to endorse any primitive relation expressions, and regard all such expressions as defined in terms of one-place predicates alone. This view is taken to be a formal counterpart of the so-called 'doctrine of internal relations': whenever two individuals x and y enter a certain relation R, there must exist certain 'internal' properties of x and of y due to which the relation R obtains. Thus, if 'hot', 'tepid', and 'cold' were the only noteworthy temperature-qualities, one might define 'x is warmer than y' by the words 'either x is hot and y is tepid, or x is hot and y is cold, or x is tepid and y is cold'. Assuming that the number of noteworthy temperature qualities is finite, the relation expression 'is warmer than' might seem to be definable in terms of one-place predicates, and hence the problem of interpreting the relation expression could be reduced to the simpler problem of interpreting predicates.

Although the metaphysical doctrine corresponding to this proposal is completely out of fashion, it is not altogether obvious why it is untenable. Of course, a theory comprising relation expressions is not 'reducible' to one containing only one-place predicates in any sense which makes it relevant to observe that a predicate calculus comprising even one relation expression has strictly greater deductive and expressive power than the monadic predicate calculus. However, apart from the fact that nominalists endorse certain (non-schematic) relation expressions (such as the symbol '$\leqslant$' of the calculus of individuals), they might not need the additional power of a full predicate calculus with schematic relation expressions. Again, it has been urged (e.g. by Carnap[16]) that it is in the interest of science to replace classificatory concepts (which can be expressed by one-place predicates) in favor of comparative ones (which are expressed by relation expressions). Thus, the comparative concept 'warmer than' enables one to express finer temperature gradations than the classificatory concepts 'hot', 'tepid', and 'cold'. Still, only finitely many temperature gradations can be experimentally distinguished. Hence, it would seem as if the entire experimentally significant content of the comparative concept can be expressed by finitely many classificatory concepts. To do so would be cumbersome. But that should not bother nominalists who make it their aim to achieve ontological economy rather than elegance of descriptions.

Nevertheless, there appear to be good reasons for rejecting the present proposal. One of them is that nominalists should allow for the logical possibility that the universe of individuals is infinite and that there really might be infinitely many gradations, say, of temperature. Thus, it is at least logically possible that statements like 'for every object x there exists an object y such that y is warmer than x' are true. And states of affairs which can possibly obtain should be describable. But it is not possible to express, in a monadic predicate calculus, statements whose truth implies that there are infinitely many objects in the universe. Furthermore, even if the number of objects and of temperature gradations would always remain finite in every universe of individuals, it would be unreasonable to impose a finite upper bound which must not be exceeded by the cardinality of any universe. And if no such bound is imposed, one cannot select any one number **n** of classificatory temperature concepts of such a sort that the relation 'warmer than' is always expressible, with respect to every finite universe, just in terms of those **n** classifications. Hence, there appears to be a genuine need for primitive schematic relation expressions.

(5) When one interprets relation expressions, order must somehow be taken into account. Since nominalists have difficulties with order which is internal to the designata of such expressions, one might attempt instead to build order into the relation of designation itself, and employ some notion of ordered designation. Thus, the expression 'is brighter than' might enter a relation of designation which assigns to it, in a sequential manner, the sun and the moon. This suggestion appears to come to this: one employs a triadic relation *Des* such that *Des* ('is brighter than', x, y) shall obtain, intuitively, whenever x is brighter than y. We shall explore, in detail, a variant of this proposal which is described as follows:

(6) We have found reasons for rejecting the position (4), mentioned above, according to which relation expressions (like 'is warmer than') should be defined in terms of one-place predicates (like 'hot', 'tepid', and 'cold'). However, there is another way which permits us to construe relations as properties. Roughly speaking, it consists in taking sentences like 'B is warmer than A' to assert of B the property of being warmer than A.

If this approach is followed up, it must be possible to interpret the quasi-predicate 'is warmer than A' in a manner acceptable to nominalists. Recalling our discussion from the previous section, we have attributed the greatest *prima facie* plausibility to that nominalistic proposal according to which every predicate should designate exactly one individual. By extending this view to the present context, we arrive at the following proposal: A relation of designation should be employed which assigns to every pair $\langle$'is warmer than', $x\rangle$, where x is an individual, that individual y which is thought of as a quale corresponding to, or an exemplar representative of, the property 'being warmer than x'.

Note that we are not recommending that relation expressions should be replaced by one-place predicates. The language to be interpreted need comprise no one-place predicates at all; in particular, it need not contain predicates like 'is warmer than A'. The construing of relations as properties now at issue does not require any language revision and does not amount to the claim that formal systems containing relation expressions are reducible to ones containing only one-place predicates. The construing in question takes place only in the semantics of a language.

Operation expressions of positive degree can be treated in an analogous manner. For example, the expression 'the father of' can be interpreted by a relation of designation which assigns to every pair $\langle$'the father of', $x\rangle$, where x is an individual in the universe, a certain individual y in that universe which is understood to be the father of x.

In the next section, we shall render the present proposal formally precise, and in Chapter 3 its adequacy will be demonstrated by constructing some formal systems which are both sound and complete relative to interpretations which accord with the informal remarks which have been made so far.

3.5. DENOTATIONLESS SYMBOLS AND NOMINALISTIC MODELS

In preceding discussions, we have repeatedly found it helpful to consider the manner in which various types of expressions are interpreted by the 'classical models' (see Definition 1.3.2.12), which are commonly employed in specifying the semantics of formal systems. Invariably, the construction of such models was found to accord with realism rather than with nominalism. But nominalists should also be able to avail themselves of model-theoretic methods. For this reason, a new notion of a model will presently be proposed which is called a 'nominalistic model' in contrast with a 'classical model'. Nominalistic models should interpret predicates and operation symbols in a manner which agrees with one of the nominalistic proposals informally discussed in the preceding sections. Furthermore, allowance must be made in such models for denotationless terms and predicates, which we shall presently consider.

In ordinary discourse there occur simple names, like 'Pegasus', compound names, like '1/0', and predicates, like 'is a winged horse', which do not apply to anything. If sentences of ordinary discourse containing such expressions are to have plausible translations among the sentences of a formal system, then the interpretation of that system should allow for some formal counterpart of the distinction between denoting and non-denoting symbols.

Furthermore, it does not appear to be true on logical grounds alone that the universe is non-empty. Hence, models should be admitted, as 'logically possible worlds', whose universe is empty. But with respect to an empty universe, all names (and expressions treated like names) must lack denotation. This then is a second reason for giving consideration to symbolic names which fail to name.

Let us briefly summarize and discuss the main types of methods which are currently employed in interpreting counterparts of denotationless terms in formal systems:

(1) There is the method, originated by Frege, of selecting some arbitrary item in the universe of discourse and translating all denotationless names of ordinary discourse into symbolic names whose common designatum is that item. Thus, if the number zero happens to belong to the given universe, one might decide to translate the names 'Pegasus' and 'Zeus' into symbolic names (individual constants) both of which denote zero. This leads to counter-intuitive translations. For example, formal counterparts of the sentences 'Pegasus = Zeus' and 'Pegasus is less than one' will turn out to be true. Furthermore, empty universes cannot be accommodated by this method.

(2) A different method, advocated by Russell and Quine, consists in construing all names (and, in particular, non-denoting ones) as concealed definite descriptions, to be eliminated in context. Thus, 'Pegasus' may be understood as 'the winged horse captured by Bellerophon' or, with some artificiality, as 'the unique object which pegasizes'; and sentences like

'Pegasus is winged' pass into false translations like 'there is one and only one item which pegasizes, and it is winged'. Since we refuse to treat predicates as syncategorematic expressions and side instead with those nominalists who treat predicates as expressions which designate individuals, this approach will not work for us. For, having eliminated the proper name 'Pegasus' in favor of logical words together with the predicate 'pegasizes', one still has to decide, according to the nominalists in question, what the predicate 'pegasizes' shall designate. Unlike realists, nominalists must account not only for denotationless expressions of the syntactical category 'names', but also for denotationless predicates.

(3) A method, favored, e.g., by Routley[17], consists in assigning to the formal counterparts of denotationless terms of English certain items which are not members of the universe (of the given model), but are instead either members of the universe of some other model (in which case one thinks of them as 'possible' items, like Pegasus), or else fail to be members of any universe (in which case they are regarded as 'impossible' items, like the round square). In trying to identify in formal contexts that relation which shall count as the designation relation, it is plausible to adopt this principle: whenever a value is assigned to an expression (in a model), that value is an object designated by the expression (in the model). Accordingly, the present approach consists in translating terms which fail to denote into terms which denote 'possible' or 'impossible' items. Since nominalists avoid reference to any items other than actual individuals, they cannot consistently avail themselves of this technique.

(4) Instead, we shall favor a method which consists in treating formal counterparts of non-denoting English expressions again simply as non-denoting. That means, roughly, that the functions in models which assign certain items to expressions shall not be defined on those expressions which suitably translate denotationless terms of ordinary discourse.

The informal discussions of this chapter have prepared us for the following definition:

(3.5.1) DEFINITION: Suppose that $\langle V, L \rangle$ is a formal system. Then **M** is a *nominalistic model* for $\langle V, L \rangle$ if and only if there exist **R, U, O, P,** and **A** such that $\mathbf{M} = \langle \mathbf{R, U, O, P, A} \rangle$, and the following conditions are satisfied:
(1) **U** is an atomistic universe of individuals for **R**,

(2) **O** is a function such that
  (a) the domain of **O** is included in the set of all $(n+1)$-term sequences $\langle o, x_0, \ldots, x_{n-1} \rangle$, where o is an **n**-place operation symbol in V and $x_0, \ldots, x_{n-1}$ are members of **U**,
  (b) the range of **O** is a subset of **U**;
(3) **P** is a function such that
  (a) the domain of **P** is included in the set of all $(n+1)$-term sequences $\langle \pi, x_0, \ldots, x_{n-1} \rangle$, where π is an $(n+1)$-place predicate in V and $x_0, \ldots, x_{n-1}$ are members of **U**,
  (b) the range of **P** is a subset of **U**;
(4) **A** is a function such that
  (a) the domain of **A** is included in the set of variables, and
  (b) the range of **A** is a subset of **U**.

The intuitive ideas carried to this definition are roughly the following:

The universe **U** of each model is a possible world of individuals which are composed of atomic parts relative to the part-whole relation **R**. (We remind of Definition 2.3.11.) The universe **U** may be empty. Atomistic universes of individuals, rather than non-atomistic ones, were chosen only because they are slightly more convenient to work with.

The constituent **P** of a nominalistic model is a function which interprets the predicates of the system in accordance with the last nominalistic proposal (6) informally introduced in the preceding section. If the one-place predicate 'is white' were to occur in the vocabulary of a system, **P** might assign to the one-term sequence $\langle\text{'is white'}\rangle$ a certain individual in the universe **U** of the model. Depending upon the system to be interpreted, that individual designatum can be thought of as the sum of all white things or as an individual quale or exemplar of the color 'white'. It could be that the predicate 'is white' should not apply to any item in the universe. In that event, the predicate is regarded as denotationless; that means, formally, that the sequence $\langle\text{'is white'}\rangle$ is not a member of the domain of **P**. A relation expression like 'is brighter than' is interpreted by **P** as follows: whenever some individual x in **U** could be said to bear the relation 'brighter than' to some individual or other, **P** will assign to the two-term sequence ⟨'is brighter than', x⟩ a certain individual y in the universe **U** which is thought of as exemplifying what it means to be less bright than x. If some individual x happens to be least bright in the universe (that is, if x is such that there exists no y in **U** such that x is brighter than y), then the particular sequence ⟨'is brighter than', x⟩ will fail to be in the domain of **P**. This means that the relation expression 'is brighter than' is not defined on x. In the event that nothing in the universe enters the relation 'brighter than', all sequences of the form ⟨'is brighter than', x⟩ will fail to be in the domain of **P**. Relation expressions of arbitrary degree **n** are similarly interpreted by construing them, in effect, as properties of the first (**n**−1) relata. Zero-place predicates (that is, sentence letters) are not admitted, since their interpretation by this method would be unnatural and unacceptable to a nominalist.

The constituent **O** of such a model interprets operation expressions analogously. Thus, if the zero-place operation expression 'Adam' should occur in the vocabulary of a formal system, and if it should be regarded as a denoting name, then **O** will assign to the one-term sequence ⟨'Adam'⟩ a certain individual in the universe **U**. If 'Adam' is considered to be a non-denoting name, then ⟨'Adam'⟩ should fail to be in the domain of **O**. If Adam is thought to have a father, then the two-term sequence ⟨'the father of', Adam⟩ should be in the domain of **O**, and **O** should assign to that sequence a certain individual in **U** which is thought of as the father of Adam. If the compound name 'the father of Adam' is regarded as denotationless, then ⟨'the father of', Adam⟩ should not be in the domain of **O**. Note that 'Adam' may be treated as a denoting name, even though 'the father of Adam' may lack denotation. As we shall show more explicitly later on, the converse cannot happen; that is, under a natural definition of 'denotation' in models it should not happen that the compound name 'the father of Adam' denotes some individual, while 'Adam' fails to denote.

The last constituent, **A**, of a nominalistic model interprets certain variables (those which have values) exactly as it was explained in connection with the models (defined by 2.5.3.1) of the calculus CII.

The values of every one of the assignments **O**, **P**, and **A** in nominalistic models are always individuals in the universe **U**; and the members of the universe **U** are always individuals in the sense explained in Chapter 2. For these reasons, the models seem to meet nominalistic requirements.

The intuitive import of nominalistic models will be further elaborated in connection with particular formal systems to be developed in the next chapter.

In addition to models which accord with nominalistic proposals, the notion of *truth* in such models (and related notions) must be so specified that it gives rise to a semantics relative to which a lower predicate calculus is adequate. However, our discussion of truth conditions in Section 3.3 indicated that the relations of predication, used in formulating truth definitions for atomic sentences, will depend on the kind of individuals to which one intends to give systematic treatment. Thus, whereas the notion of a nominalistic model could be specified quite generally, the notion of truth in such models must be defined in connection with particular systems yet to be developed.

In the next chapter, we shall construct one example from each of three general types of nominalistic theories. In trying to meet at first rather liberal and then increasingly stringent nominalistic demands, we shall first construct a theory whose individuals are sets; next, a theory whose individuals are no longer sets, though they are still abstract; and finally we shall consider a theory which is intended to treat of concrete individuals.

REFERENCES

[1] Quine (1953), p. 103.
[2] Church (1958), p. 1014, footnote.
[3] Such a theory of descriptions is formally treated in Kalish and Montague (1964), Ch. 7.
[4] Quine (1953), p. 13.
[5] Cartwright (1954).
[6] For example, in Eberle (1969b).
[7] Quine (1953), pp. 115–6.
[8] Tarski and Vaught (1957).
[9] Russell (1946).
[10] Martin (1958).
[11] Tarski (1936).
[12] Quine (1940), pp. 115–6.
[13] A systematic treatment of senses in the framework of an interpreted intensional logic is given in Kaplan (1964).
[14] In a slightly different context, this point is made by Quine (1940), pp. 72f.
[15] The identification of nominalism with a resemblance theory of classification is made, e.g., in Bochenski (1956).
[16] Carnap (1966), Chapter 5.
[17] Routley (1966).

CHAPTER 4

BUNDLES OF QUALITIES, QUALITIES, AND CONCRETA

In the preceding chapter, the general notion of a nominalistic model appropriate to first-order systems has been discussed and formulated, and informal consideration was given to truth conditions which might be specified for the sentences of such systems. The present chapter is in the main devoted to the formulation and discussion of some particular formal systems containing predicates and operation symbols. These systems are to be interpreted by means of nominalistic models and by the use of truth conditions which illustrate the general pattern of such conditions informally discussed in the preceding chapter. The first of these interpreted formal systems exemplifies the position according to which individuals may be construed as sets. In an extension of that system, axiomatic treatment is given to a notion of resemblance. The second system is to be an instance of the position that individuals may be abstract items, though not sets. Finally, the semantics of a third system, only presented in outline, is regarded as a step toward the reconstruction of that nominalistic position according to which all individuals are concrete.

4.1. BUNDLES OF QUALITIES AND THEIR CALCULUS CB

4.1.1. *Background*

The view that all things in the empirical world are just 'bundles' or 'combinations' of qualities is one which has been advocated at least since the time of Berkeley by many empirically minded philosophers. Among recent proponents of this view, Russell is probably best known.[1] Without attempting either to trace the history of this position or to give anything like a complete description of its tenets, a few remarks are made to remind the reader of the doctrine in question.

It is doubtful whether either proponents or opponents of this view have ever stated precisely what is meant by a 'bundle of qualities'. One learns that certain qualities are 'in' such bundles, that smaller bundles may be 'in' larger ones, and that two such bundles differ just in case the qualities 'in' the bundles differ.

Like the notion of a 'bundle', so also that of a 'quality' is in need of clarification. As a principle of individuation concerning the qualities in bundles, the following is sometimes offered: two such qualities are identical just in case they resemble (or, perhaps, are indistinguishable from) the same qualities. The intended qualities are usually thought to be positive rather than negative, simple rather than compound or complex, natural rather than artificial, occurrent rather than dispositional, and often also phenomenal rather than physical. Favorite constituents of 'bundles' are such qualities as specific phenomenal shades of color, sounds of determinate pitch and loudness, specific tactile presentations, odors, savors, moments in subjective time, and locations in the field of vision.

Varied support is claimed for the view that things in the world are just bundles of qualities. Often reasons are cited for rejecting a certain metaphysical doctrine of substances, according to which each object contains a material substratum whose only function it is to underlie or support all qualities of the object, while being distinct from and not constituted of those qualities. The assumption of such a substance, it is felt, is either incomprehensible or superfluous: an object not only has qualities; it is rather constituted by its qualities. In addition, epistemological considerations are often brought into play: the objects of immediate experience are certain qualities themselves, and not substances which might support these qualities. Distinct particulars, it is sometimes held, must noticeably differ with respect to some qualities, and it is only qualities which are directly noticeable. Hence, the principle of identity appropriate to empirical things must be formulated in terms of their noticeable qualities, and not in tems of some theoretical concept of substance which bears some non-empirical relation other than a part-whole relation to its noticeable qualities.

We do not make it our task to examine historically given views or the support alleged for such views in regard to 'bundles of qualities'. Accordingly, we cannot claim that the system about to be developed has a specific historic precedent whose reconstruction it is meant to be. It is rather advanced as a proposal which was stimulated by and has some affinities to the historical doctrine in question. Still, in order to draw attention to these similarities, we shall call the system 'CB', short for 'calculus of bundles'.

In formulating the calculus CB, we follow again the pattern set in earlier chapters. We shall describe, in the next section, the vocabulary of CB (which is not essential to the philosophical train of thought); and subsequently we shall present its semantics, interlaced with extensive informal explanations.

4.1.2. *The Vocabulary of CB**

The vocabulary V of the formal system CB comprises the following categories of symbols:

(1) One *primitive logical constant*: the two-place non-schematic predicate 'M', where formulas of the form 'xMy' are read 'x matches y'. This constant will have precisely the same properties as did the constant '∘' of overlapping in the calculus CII. The change of notation and of the informal reading are dictated solely by the intuitive interpretation which we shall wish to carry to the notion of overlapping in the present development.

(2) The four *defined* non-schematic predicates of CII:

(a) The two-place predicate '$\leqslant$'. Formulas of the form '$x \leqslant y$' are read 'x is part of y'.

(b) The two-place predicate '$<$'. Formulas of the form '$x<y$' are read 'x is a proper part of y'.

(c) The two-place predicate '$=$'. Formulas of the form '$x=y$' are read 'x is the same as y'.

(d) The one-place predicate 'At'. Formulas of the form 'Atx' are read 'x is atomic' or 'x is simple'.

(3) For every natural number **n**, there is a denumerable set of **n**-place *operation symbols*.

(4) For every positive integer **n**, there is a denumerable set of **n**-place *schematic predicates*.

In addition, CB, like every system, has a denumerable set of individual variables.

4.1.3. *The Semantics of CB with Informal Discussion*

In conformity with previous remarks, the interpretation of CB is to be given by specifying a set of nominalistic models relative to some appropriate part-whole relation. We first consider what conditions should be imposed on part-whole relations if the interpretation of CB is to be in intuitive accord with the view that all individuals are bundles of qualities.

A 'bundle' or 'collection' or 'combination' of qualities, we submit, can be regarded as a set of qualities in some universe of individuals; that is, a 'bundle' is construed as a *set which is also an individual*. This understanding of what is meant by a 'bundle' is supported by the principle of individuation held to be appropriate to bundles of qualities: two bundles of qualities are said to differ just in case there is some quality 'in' one bundle which is not 'in' the other. If bundles are regarded as sets and 'in' is understood in the sense of membership, then the formal counterpart of this principle appears to be just the principle of extensionality. Further, if two bundles of qualities in the empirical world differ just with respect to their spatial or temporal positions, this is taken to imply that these bundles comprise different qualities (position qualities). Thus, 'bundles', in this context, are not taken to be 'configurations' or 'patterns' of qualities; they are taken to differ with respect to the nature or number, not with respect to the arrangement of their constituents. In this respect also, bundles of qualities are like sets of qualities.

Whereas the notion of a 'bundle' had to be explicated, that of a 'quality' may be left on the intuitive level for the moment.

With reference to the Definition (3.5.1) of 'nominalistic model', the models of CB can be easily characterized as follows:

(4.1.3.1) DEFINITION: **M** is a *model for* CB if and only if for some **R, U, O, P**, and **A**, **M** = ⟨**R, U, O, P, A**⟩, **M** is a nominalistic model, and the part-whole relation **R** is the set-theoretic inclusion relation confined to some set.

Recall that by the definition of 'nominalistic model', the constituent **U** of such models is an atomistic universe of individuals for **R** (see Definition 2.3.11), which implies that **R** is a part-whole relation (see Definition 2.2.6). Hence, for every relation **R** in a model of CB there will exist an infinite set *A* of items which are least, in the field of **R**, with respect to inclusion, and all items which enter the relation **R** will be unions (in the set-theoretic sense) of non-empty subsets of A.

We shall presently introduce, by recursive definitions, the notions of denotation and of satisfaction in models of CB, and subsequently illustrate and discuss these formal goings-on.

(4.1.3.2) DEFINITION: $Den(\mathbf{M}, \tau, x)$ [read: in **M**, τ denotes x] if and only if for some **R, U, O, P**, and **A**, $\mathbf{M} = \langle \mathbf{R, U, O, P, A} \rangle$ is a model of CB, and either:
(1) τ is a variable in the domain of **A** and $x = A(\tau)$, or
(2) for some natural number **n**, for some **n**-place operation symbol δ in the vocabulary of CB and for some sequence $\langle \zeta_0, \ldots, \zeta_{n-1} \rangle$ of terms of CB, $\tau = \ulcorner \delta \zeta_0 \ldots \zeta_{n-1} \urcorner$ and there exists a sequence $\langle y_0, \ldots, y_{n-1} \rangle$ such that the (**n**+1)-term sequence $\langle \delta, y_0, \ldots, y_{n-1} \rangle$ is in the domain of **O**, $x = \mathbf{O}(\langle \delta, y_0, \ldots, y_{n-1} \rangle)$, and for each $i < \mathbf{n}$, $Den(\mathbf{M}, \zeta_i, y_i)$.

The notation 'A_x^α' will continue to refer to that assignment which differs from A, if at all, only by assigning x to α (see Definition 2.5.3.2).

(4.1.3.3) DEFINITION: **M** sat ϕ [read: **M** satisfies ϕ] if and only if for some **R, U, O, P**, and **A**, $\mathbf{M} = \langle \mathbf{R, U, O, P, A} \rangle$ is a model of CB, and either:
(1) for some terms τ_0, τ_1 of CB, $\phi = \ulcorner \tau_0 M \tau_1 \urcorner$, for some x_0 and x_1 $Den(\mathbf{M}, \tau_0, x_0)$ and $Den(\mathbf{M}, \tau_1, x_1)$, and there exists a y in **U** such that y bears **R** [i.e. $\subseteq$] to both x_0 and x_1;
(2) for some positive integer **n**, for some **n**-place predicate π in the vocabulary of CB and for some **n**-term sequence $\langle \tau_0, \ldots, \tau_{n-1} \rangle$ of terms of CB, $\phi = \ulcorner \pi \tau_0 \ldots \tau_{n-1} \urcorner$ and there is a sequence $\langle x_0, \ldots, x_{n-1} \rangle$ such that for each $0 < i < n$ $Den(\mathbf{M}, \tau_i, x_i)$, the sequence $\langle \pi, x_0, \ldots, x_{n-2} \rangle$ is in the domain of **P**, and there exists a y in **U** such that y bears **R** [i.e. $\subseteq$] to both x_{n-1} and to $\mathbf{P}(\langle \pi, x_0, \ldots, x_{n-2} \rangle)$;
(3) for some formula ψ, $\phi = \ulcorner \sim \psi \urcorner$ and it is not the case that **M** sat ψ;
(4) for some formulas ψ and χ, $\phi = \ulcorner (\psi \rightarrow \chi) \urcorner$ and **M** sat ψ only if **M** sat χ; or

BUNDLES OF QUALITIES, QUALITIES, AND CONCRETA   153

(5) for some formula ψ and variable α, $\phi = \ulcorner \forall \alpha \psi \urcorner$ and for every x in $\mathbf{U}$, $\langle \mathbf{R}, \mathbf{U}, \mathbf{O}, \mathbf{P}, \mathbf{A}^\alpha_x \rangle$ sat ψ.

(4.1.3.4) DEFINITION: ϕ is *true in* $\mathbf{M}$ if and only if ϕ is a sentence of CB and $\mathbf{M}$ sat ϕ.

Let us note, to begin with, that according to Definition (4.1.3.2) a compound term denotes something only if every simpler term occurring in it denotes something; and that both simple and compound terms may lack denotation. Thus, every variable which is not in the domain of the constituent **A** in a model, will fail to denote in that model. Similarly, if δ is a zero-place operation symbol [that is, a constant or a symbolic name] then, according to clause (2) of that definition, δ may fail to denote if the one-term sequence $\langle \delta \rangle$ is not in the domain of the assignment **O**.

We understand, in this connection, that the sequence $\langle \delta, \zeta_0, ..., \zeta_{0-1} \rangle$ is the concatenate of the sequences $\langle \delta \rangle$ and $\langle \zeta_0, ..., \zeta_{0-1} \rangle$, which is in turn the concatenate of $\langle \delta \rangle$ and the empty set 0, which is just $\langle \delta \rangle$.

A compound term of the form '$\delta \zeta$', composed of a one-place operation symbol δ [say, 'the tail of'] and a zero-place operation symbol ζ [say, 'Hydra'] may fail to denote even if the name ζ denotes something, say x. This can happen if the sequence $\langle \zeta, x \rangle$ fails to be in the domain of the assignment **O**. On the other hand, if the simple name ζ fails to denote, then the compound name '$\delta \zeta$' is bound to remain denotationless. Thus, if 'Hydra' should be a denotationless name, then the compound name 'the tail of Hydra' must also lack denotation.

The principle, that compound terms shall denote only if simpler terms occurring in them denote, is quite plausible with respect to extensional terms, but not with respect to ones which introduce non-extensional contexts, like 'the picture of Hydra'. For the picture of Hydra may well exist even when Hydra does not. However, we shall not, in our present extensional logic, be concerned with referentially opaque compound terms. Generally, non-extensional contexts present considerable difficulties even to realists, and we shall not attempt to meet, in the present work, the additional difficulties which arise from such contexts for nominalists.

Consider next the intuitive import of the satisfaction clause (2) in (4.1.3.3). We omit all reference to models and replace satisfaction by truth whenever doing so contributes to easier readings. Suppose that π is a one-place predicate, say 'is white', and that τ is a name, say 'the Taj Mahal', so that "$\pi\tau$" corresponds to the sentence 'the Taj Mahal is white'. Informally, the following is a satisfaction condition in agreement with clause (2):

> 'The Taj Mahal is white' is a true sentence just in case there is a sequence $\langle x \rangle$ such that 'the Taj Mahal' denotes x, the sequence $\langle$'is white'$\rangle$ is interpreted by the referential assignment **P** of the model, and some bundle y is included both in the designatum x of 'the Taj Mahal' and in the designatum $\mathbf{P}(\langle$'is white'$\rangle)$.

Thus, intuitively, the bundle designated (according to **P**) by the predicate 'is white' should share an element with all and only those individual bundles which are white. Now, the only kinds of bundles which meet these intuitive requirements are ones which one would call bundles of 'qualities'. For, a bundle composed of just those shades of the color 'white' which deserve to be called 'white' satisfies the intuitive requirement, while a bundle comprising anything other than qualities or anything other than shades of white would intuitively violate the formal truth condition. Thus, while we have not actually defined the notion of a quality, the system as a whole is so interpreted that it seems congenial to no extra-systematic description other than the one according to which the system treats of bundles of qualities.

Note that sentences like '$\pi\tau$' can be true, according to clause (2) of (4.1.3.3), only if the term τ denotes something. Thus, if the name 'Pegasus' fails to denote, then the atomic sentence 'Pegasus is white' will be false. 'Pegasus is so-and-so' may appear to be a true sentence if the predicate 'so-and-so' is replaced by one which makes the sentence analytic. Thus, 'Pegasus is winged' may seem to be true, contrary to our formal definition, even if 'Pegasus' fails to denote. However, if an adequate treatment of non-extensional contexts were contemplated, it would seem reasonable to interpret analyticity or necessity in this connection as follows:

> 'Pegasus is winged' is necessarily true just in case for every possible world (or model): if 'Pegasus' denotes something in that world, then 'Pegasus is winged' is true of that world.

This interpretation is quite compatible with the assumptions that 'Pegasus' fails to denote in the actual world and that 'Pegasus is winged' is false of that world. Again, the sentence 'Pegasus is so-and-so' may appear true, contrary to our definition, if the predicate 'so-and-so' expresses some negative quality or so-called 'privation'. Thus 'Pegasus fails to be white' seems true. However, we take the traditional doctrine regarding bundles of qualities to imply that the qualities in such bundles are always positive ones. Thus, 'failing to be white' would not be regarded as a suitable constituent of bundles of qualities. In accord with the customary notion of an 'explicit translation', those predicate phrases of ordinary discourse which are taken to express negative qualities pass into explicit translations which contain negation signs, and not into atomic sentences. Accordingly, we regard only those predicate phrases of ordinary discourse as appropriate translations of our formal predicates which in that discourse are taken to express simple positive qualities.

While our interpretation of atomic sentences seems to capture most of the content traditionally carried by the notion of a bundle of qualities, the epistemological import of that doctrine is left out of account. Nothing in our formal development suggests that the qualities in our bundles should be phenomenal ones or otherwise epistemologically distinguished. Accordingly, we regard our theory as treating of bundles of qualities, but not necessarily of bundles comprising phenomenal qualities or 'ideas'.

Still with reference to clause (2) of Definition (4.1.3.3), consider how truth conditions are specified for atomic sentences of relational form. Assume that π is a two-place predicate, that τ_0 and τ_1 are names denoting, respectively, the moon and the sun and that the sentence '$\pi\tau_0\tau_1$' is translated into 'the moon is less bright than the sun'. Then the following is an informal example of the second clause in our truth definition:

> The sentence 'the moon is less bright than the sun' is true just in case there is a two-term sequence $\langle x_0, x_1 \rangle$ [say, the sequence ⟨the moon, the sun⟩] such that 'the moon' designates the moon, 'the sun' designates the sun, the sequence ⟨'is less bright than', the moon⟩ is interpreted by **P**, and there exists a bundle y which is included both in the sun and in that bundle which is assigned to ⟨'is less bright than', the moon⟩.

The bundle of qualities assigned to the sequence ⟨'is less bright than', the moon⟩ should intuitively be that bundle which might qualify as the designatum of the one-place predicate 'is brighter than the moon'. Thus, intuitively, one seeks a bundle which shares an element with all and only those other individuals which are brighter than the moon. The bundle in question should comprise qualities; in fact, just those qualities which characterize individuals brighter than the moon. Again, although we have not formally defined the notion of a quality, no examples of such bundles come to mind other than ones which comprise qualities. However, there is no need to decide what kind of qualities should go into the bundle designated by 'is brighter than the moon': one might think of this bundle as comprising a single quality, that of being brighter than the moon, or perhaps a number of qualities corresponding to positions on some brightness scale.

If the moon should be the brightest object, or completely incomparable with respect to brightness, then the predicate 'is brighter than the moon' would be regarded as denotationless, and accordingly, the pair ⟨'is less bright than', the moon⟩ would not be in the domain of **P**; just as, in the earlier example, the one place predicate 'is white' might fail to designate altogether by virtue of the possibility that ⟨'is white'⟩ fails to be in the domain of **P**.

As in the case of one-place predicates, atomic sentences like '$\pi\tau_0\tau_1$' containing a two-place predicate π will automatically be false if one of the terms τ_0 or τ_1 fails to denote anything. Thus, formal counterparts of the sentence 'Pegasus is heavier than Homer' would be false just on the grounds that the name 'Pegasus' is denotationless. Just as all simple predicates are taken to express simple positive qualities, so also all simple relation expressions stand for simple positive relations. The justification is analogous.

It will appear already, that a general substitution rule on predicates cannot hold of our system. Thus, '$\sim\pi\tau$' or '$\sim\pi\tau_0\tau_1$' cannot be regarded, as they can in conventional systems, as instances of the respective sentences '$\pi\tau$' and '$\pi\tau_0\tau_1$', since the latter sentences will be false if one of the τ's lacks denotation, while the former sentences will be true. Hence, our predicates deserve to be called 'schematic' only in the sense that one may replace such predicates by any other (simple) predicates of the same grammatical status.

Turning now to clause (1) in the Definition (4.1.3.3) of satisfaction,

BUNDLES OF QUALITIES, QUALITIES, AND CONCRETA 157

we note that the interpretation given to the formula 'xMy' is just the same as that given, in connection with CII to the formula '$x \circ y$' (in verifying this, it helps to recall Remark 2.5.3.4). Hence, the formula 'xMy' [read: 'x matches y'] differs only notationally from '$x \circ y$' [read: 'x overlaps y']. Since individuals are presently regarded as bundles of qualities, overlapping individuals will now be individuals which have a common quality, and this is at least one of the meanings of the word 'matching'. In ordinary discourse, e.g. two red objects are said to 'match' or to 'resemble' one another either if they have exactly the same shade of color, or if they have shades of color which are distinct but sufficiently close to fall within some narrow range of colors. The first sense of 'matching' may be called that of 'partial qualitative identity', whereas the latter sense may be expressed by 'partial qualitative similarity'. Matching is here construed in the former sense. Partial qualitative similarity will be called 'resemblance', in distinction to 'matching', and this notion will be examined later.

Our present interpretation of 'matching' is an extension of the previous treatment of 'overlapping' in so far as terms other than variables are taken into account. As in the case of variables, it will hold regarding arbitrary terms τ_0 and τ_1 that formulas of the form '$\tau_0 M \tau_1$' are satisfied only if τ_0 and τ_1 are denoting terms. Expressions of reflexivity '$\tau M \tau$' will be true if and only if τ is a denoting term and hence serve to express the same content as the sentence 'τ exists'.

By combining the satisfaction conditions (1) and (2) of (4.1.3.3), we obtain the following informal example of a derivative truth condition:

> If τ_0 is any name, and τ_1 is a name which denotes the designatum, according to P, of $\langle$'is white'$\rangle$, then the sentence 'τ_0 is white' is true just in case the sentence 'τ_0 matches τ_1' is true.

Thus, one regards as white all and only those individuals which match the designatum of the predicate 'is white'. The relation of predication employed in CB is therefore just the relation of matching, which accords with the preliminary discussion of Section 3.3.

All other semantical notions of CB are as expected. In particular, the notions ϕ *is valid* (ϕ is satisfied by all models) and *K semantically yields* ϕ (every model which satisfies all members of K also satisfies ϕ)

remain just as previously defined in connection with CII (Definitions 2.5.3.5 and 2.5.3.6), except that all reference to models is appropriately understood as a reference to the models of CB.

4.1.4. *The Axioms of CB*

In reading the subsequent formulas, assume throughout that **n** is a natural number, that ζ and η are terms, that $\langle \zeta_0, ..., \zeta_{n-1} \rangle$ is an **n**-term sequence of terms, that α, β, and γ are variables and $\langle \alpha_0, ..., \alpha_{n-1} \rangle$ an **n**-term sequence of variables all of which are distinct, that β does not occur in either ζ or η, that δ is an **n**-place operation symbol, that π is an **n**-place predicate provided **n** is positive, that ϕ is a formula, and that ϕ_ζ^α is the proper substitution of the term ζ for the variable α throughout the formula ϕ.

With minor notational differences, the *definitional axioms* of CB are just those of CII. For convenient reference, we state them again, together with informal readings:

(D1) $\qquad \zeta \leqslant \eta \leftrightarrow [\zeta \operatorname{M} \zeta \ \& \ (\forall \beta)(\beta \operatorname{M} \zeta \to \beta \operatorname{M} \eta)]$.

In words: ζ is part of η just in case ζ matches itself (or, ζ exists) and every bundle which matches ζ matches η.

(D2) $\qquad \zeta < \eta \leftrightarrow (\zeta \leqslant \eta \ \& \sim \eta \leqslant \zeta)$.

In words: ζ is a proper part of η just in case ζ is part of η, but η is not part of ζ.

(D3) $\qquad At\,\zeta \leftrightarrow [\zeta \operatorname{M} \zeta \ \& \sim (\exists \beta)(\beta < \zeta)]$.

In words: ζ is an atom (or simple) just in case ζ matches itself (or, ζ exists) and ζ has no proper part.

(D4) $\qquad \zeta = \eta \leftrightarrow (\zeta \leqslant \eta \ \& \ \eta \leqslant \zeta)$.

In words: Two individuals are identical just in case each is part of the other.

The *proper axioms* include all proper axioms of CII, which were:

(Ax1) $\qquad [\zeta \operatorname{M} \zeta \ \& \ (\forall \alpha)\,\phi] \to \phi_\zeta^\alpha$.

In words: If ζ matches itself (or, if ζ exists) and if everything satisfies the condition ϕ, then so does ζ.

(Ax2) $(\forall \alpha)\, \alpha\, M\, \alpha$.

In words: every actual individual matches itself (or, every actual individual exists).

(Ax3) $\zeta\, M\, \eta \leftrightarrow (\exists \beta)(At\, \beta\, \&\, \beta \leqslant \zeta\, \&\, \beta \leqslant \eta)$.

In words: two bundles ζ and η match just in case they have an atomic bundle as a common part.

In addition, the following proper axioms are peculiar to CB:

(Ax4) $\delta \zeta_0 \ldots \zeta_{n-1}\, M\, \delta \zeta_0 \ldots \zeta_{n-1} \to [\zeta_0\, M\, \zeta_0\, \&\, \cdots\, \&\, \zeta_{n-1}\, M\, \zeta_{n-1}]$.

In words: Suppose that $\delta \zeta_0 \ldots \zeta_{n-1}$ is a compound term formed by applying the **n**-term operation symbol δ to the **n** simpler terms $\zeta_0, \ldots, \zeta_{n-1}$, then the compound term has denotation [i.e. denotes something which matches itself] only if every simpler term occurring in it has denotation. Or, the result of applying the operation δ to all of the ζ's exists only if each of the ζ's exists.

(Ax5) $\pi \zeta_0 \ldots \zeta_{n-1} \to [\zeta_0\, M\, \zeta_0\, \&\, \cdots\, \&\, \zeta_{n-1}\, M\, \zeta_{n-1}]$.

In words: Assuming that **n** is positive, **n** individuals $\zeta_0, \ldots, \zeta_{n-1}$ enter an **n**-adic relation π only if each of the individuals ζ_i matches itself (or, exists). Still less formally: all simple relation expressions and predicates π express positive relations and properties.

(Ax6) Suppose that ϕ' results from ϕ by replacing one or more free occurrences of the term ζ by free occurrences of the term η. Then $(\zeta = \eta\, \&\, \phi) \to \phi'$.

In words: If ζ and η are identical and if ζ satisfies the condition ϕ, then so does η. This is just the customary principle of the indiscernability of identicals.

(Ax7) $(\exists \beta)\, \pi \alpha_0 \ldots \alpha_{n-2} \beta \to (\exists \gamma)(\forall \beta)\, [\beta\, M\, \gamma \leftrightarrow \pi \alpha_0 \ldots \alpha_{n-2} \beta]$.

In words: If the first $(n-1)$ relata of an **n**-adic relation π are related, by π, to something or other, then there exists a bundle of qualities which matches exactly those bundles to which the first $(n-1)$ relata are related by π. Or, more freely: Whenever an **n**-adic relation π obtains at all, then there exists a certain bundle of qualities (the one which is assigned to π with respect to its first $(\mathbf{n}-1)$ relata) which shares a quality with exactly those items to which the first $(\mathbf{n}-1)$ things can bear the relation π.

In the event that $\mathbf{n}=1$, it is understood that the concatenate of the sequence $\langle \alpha_0, ..., \alpha_{1-2} \rangle$ is just the empty set and that formulas of the form '$\pi\alpha_0 ... \alpha_{1-2}\beta$' which may appear in (Ax7) simply degenerate to formulas of the form '$\pi\beta$'. Thus, according to (Ax7), if a one-place predicate π applies to anything at all, then there exists a bundle of qualities (intuitively the designatum of π) which shares a quality with all and only those bundles to which the predicate π properly applies.

Since the intuitive import of the axioms (Ax4), (Ax5) and especially (Ax7) is not easily grasped upon first reading, we refer to the extensive informal discussion given in the preceding section to the corresponding semantical principles.

The *logical axioms* of CB are those of CI I.

The *theorems* of CB include all the theorems (in Section 2.5.2) of CII under our present convention regarding Greek symbols, and upon replacement of the symbol '∘' by 'M'.

As it has become our habit, we shall demonstrate, in the next section, the semantical adequacy of CB. We shall conclude, roughly speaking, that exactly those formulas are derivable from our axioms which are true in every model, or in every possible world of CB. Readers who are not interested in proofs may skip the next section.

4.1.5. *The Semantical Adequacy of CB**

In order to avoid undue repetitions, we shall frequently refer to the theorems and proofs given in Section 2.5.4 in connection with CII.

We shall presuppose that the notation 'θ^α_τ' [which is read: 'the proper substitution of the term τ for the variable α throughout the well-formed expressed θ'] has been adequately defined, in the customary manner, for all variables α, for all terms τ, and for all terms or formulas θ.

The following lemmas, which have counterparts in traditional logic, are easily shown by induction on θ, the steps in the induction being similar to the recursive conditions which appear in Definitions (4.1.3.2) and (4.1.3.3):

(4.1.5.1) LEMMA: Suppose that $\mathbf{M}=\langle \mathbf{R}, \mathbf{U}, \mathbf{O}, \mathbf{P}, \mathbf{A}\rangle$ and $\mathbf{N}=\langle \mathbf{R}, \mathbf{U}, \mathbf{O}, \mathbf{P}, \mathbf{B}\rangle$ are models of CB and that for every variable α which is free in θ (a) α is in the domain of $\mathbf{A}$ if and only if α is in the domain of $\mathbf{B}$, and (b) $A(\alpha)=B(\alpha)$. Then
(1) if θ is a term, then, for all x, Den$(\mathbf{M}, \theta, x)$ if and only if Den$(\mathbf{N}, \theta, x)$, and
(2) if θ is a formula, then $\mathbf{M}$ sat θ if and only if $\mathbf{N}$ sat θ.

(4.1.5.2) LEMMA: Suppose that $\mathbf{M}=\langle \mathbf{R}, \mathbf{U}, \mathbf{O}, \mathbf{P}, \mathbf{A}\rangle$ and $\mathbf{N}=\langle \mathbf{R}, \mathbf{U}, \mathbf{O}, \mathbf{P}, \mathbf{A}^{\alpha}_{x}\rangle$ are models of CB, that α is a variable, τ is a term, and Den$(\mathbf{M}, \tau, x)$. Then
(1) if θ is a term, then, for all y, Den$(\mathbf{M}, \theta^{\alpha}_{\tau}, y)$ if and only if Den$(\mathbf{N}, \theta, y)$, and
(2) if θ is a formula, then $\mathbf{M}$ sat θ^{α}_{τ} if and only if $\mathbf{N}$ sat θ.

(4.1.5.3) LEMMA: Suppose that $\mathbf{M}=\langle \mathbf{R}, \mathbf{U}, \mathbf{O}, \mathbf{P}, \mathbf{A}\rangle$ is a model of CB and that τ is a term. Then
(1) if Den$(\mathbf{M}, \tau, x)$ then x is a member of $\mathbf{U}$, and
(2) if Den$(\mathbf{M}, \tau, x)$ and Den$(\mathbf{M}, \tau, y)$, then $x=y$.

All of the Lemmas (2.5.4.3)–(2.5.4.9) carry over from the calculus CII with only minor modifications. In particular, the syntactical Compactness Theorem, the Deduction Theorem, and Universal Generalization with respect to a variable not free in the premise set, all continue to hold, and every consistent set can be extended to a maximally consistent set which is omega-complete in the restricted sense expressed by Lemma (2.5.4.9). The following is an extension of Lemma (2.5.4.11):

(4.1.5.4) LEMMA: Suppose that $\mathbf{M}=\langle \mathbf{R}, \mathbf{U}, \mathbf{O}, \mathbf{P}, \mathbf{A}\rangle$ is a model of CB. Then for all terms ζ and η:
(1) $\mathbf{M}$ sat $\ulcorner\zeta\leqslant\eta\urcorner$ just in case for some x and y in $\mathbf{U}$, Den$(\mathbf{M}, \zeta, x)$, Den$(\mathbf{M}, \eta, y)$ and $x\subseteq y$,
(2) $\mathbf{M}$ sat $\ulcorner\zeta<\eta\urcorner$ just in case for some x and y in $\mathbf{U}$, Den$(\mathbf{M}, \zeta, x)$, Den$(\mathbf{M}, \eta, y)$ and $x\subset y$,
(3) $\mathbf{M}$ sat $\ulcorner\zeta=\eta\urcorner$ just in case for some x in $\mathbf{U}$, Den$(\mathbf{M}, \zeta, x)$ and Den$(\mathbf{M}, \eta, x)$,
(4) $\mathbf{M}$ sat $\ulcorner At\ \zeta\urcorner$ just in case for some x in $\mathbf{U}$, Den$(\mathbf{M}, \zeta, x)$ and x is $\mathbf{R}$-least in $\mathbf{U}$ [$\mathbf{R}$ being the inclusion relation].

The proof of this lemma is similar to that given for Lemma (2.5.4.11).

(4.1.5.5) LEMMA: For every **n**-place operation symbol δ and for every **n**-term sequence $\langle \zeta_0, \ldots, \zeta_{n-1} \rangle$ of terms, the following is valid:
⌜$\delta\zeta_0 \ldots \zeta_{n-1}$ M $\delta\zeta_0 \ldots \zeta_{n-1} \to [\zeta_0$ M $\zeta_0 \& \ldots \& \zeta_{n-1}$ M $\zeta_{n-1}]$⌝.

PROOF: Assume that $M = \langle R, U, O, P, A \rangle$ is a model of CB and that **M** sat ⌜$\delta\zeta_0 \ldots \zeta_{n-1}$ M $\delta\zeta_0 \ldots \zeta_{n-1}$⌝. By definition, there is an x such that Den$(M, \ulcorner\delta\zeta_0 \ldots \zeta_{n-1}\urcorner, x)$. By Definition (4.1.3.2) Part (2), there exist $y_0, \ldots, y_{n-1}$ such that for each $i < n$, Den(M, ζ_i, y_i). By Lemma (4.1.5.3), for each such i, y_i is in U. Hence, by the definition of satisfaction, for each such i, **M** sat ⌜ζ_i M ζ_i⌝.

(4.1.5.6) LEMMA: For every **n**-place π and for every **n**-term sequence $\langle \zeta_0, \ldots, \zeta_{n-1} \rangle$ of terms, the following formula is valid:
⌜$\pi\zeta_0 \ldots \zeta_{n-1} \to [\zeta_0$ M $\zeta_0 \& \ldots \& \zeta_{n-1}$ M $\zeta_{n-1}]$⌝.

PROOF: Assume that $M = \langle R, U, O, P, A \rangle$ is a model of CB and that **M** sat ⌜$\pi\zeta_0 \ldots \zeta_{n-1}$⌝. By Definition (4.1.3.3) Part (2) for each $i < n$, Den(M, ζ_i, x_i) for some x_i. From now on the proof runs like the previous one.

(4.1.5.7) LEMMA: For every **n**-place predicate π, for every $(n-1)$-term sequence $\langle \alpha_0, \ldots, \alpha_{n-2} \rangle$ of distinct variables also distinct from the variables β and γ, the following is valid:
⌜$(\exists \beta) \pi\alpha_0 \ldots \alpha_{n-2}\beta \to (\exists \gamma)(\forall \beta) [\beta$ M $\gamma \leftrightarrow \pi\alpha_0 \ldots \alpha_{n-2}\beta]$⌝.

PROOF: Assume that $M = \langle R, U, O, P, A \rangle$ is a model of CB and that **M** sat ⌜$(\exists \beta)\pi\alpha_0 \ldots \alpha_{n-2}\beta$⌝. Hence, for some x in U, $M' = \langle R, U, O, P, A_x^\beta \rangle$ sat ⌜$\pi\alpha_0 \ldots \alpha_{n-2}\beta$⌝. Let $f = \langle \pi, A(\alpha_0), \ldots, A(\alpha_{n-2}) \rangle$ and let $p = \mathbf{P}(f)$. By the satisfaction condition (2), f is in the domain of **P**, and by Definition (3.5.1), p is in U. Assume that y is an arbitrary element of U and let $N = \langle R, U, O, P, A_{py}^{\gamma\beta} \rangle$. By Definition (4.1.3.3), Parts (1) and (2), it follows easily that **N** sat ⌜β M γ⌝ if and only if **N** sat ⌜$\pi\alpha_0 \ldots \alpha_{n-2}\beta$⌝, which essentially completes the proof.

(4.1.5.8) THEOREM (Soundness): All axioms of CB are valid, and the inference rules of CB preserve validity.

We have shown, in outline, the validity of the axioms peculiar to CB. The other proofs are similar to the ones mentioned in Section 2.4.5, except that reference to Lemma (2.5.4.2) is here replaced by an appeal to Lemma (4.1.5.2). The Indiscernability of Identicals (Ax6) is shown to

BUNDLES OF QUALITIES, QUALITIES, AND CONCRETA   163

be valid by induction on ϕ, making reference, in the inductive hypothesis, to all terms ζ and η. Although we admit non-denoting terms and empty universes, there is no difficulty (as there is in some systems) in showing that Modus Ponens preserves validity.

(4.1.5.9)   LEMMA: Suppose that K is a consistent set of formulas and that there are infinitely many variables which are not free in any member of K. Then there exists a model **M** such that **M** satisfies every member of K.

PROOF: Assume the hypothesis.

(1) By Lemma (2.5.4.9), there exists a consistent extension K' of K such that for all formulas of the form $\ulcorner(\forall\alpha)\phi\urcorner$ there is a variable β such that $\ulcorner(\phi^\alpha_\beta \vee \sim\beta \text{ M } \beta) \to (\forall\alpha)\phi\urcorner$ is in K'.

(2) By Lemma (2.5.4.8), there exists a maximally consistent extension K^* of K'.

(3) Let $At(\alpha)=$ the set of all variables β such that $\ulcorner At\,\beta\,\&\,\beta\leqslant\alpha\urcorner$ is in K^*.

(4) Let $A_0=$ the set of all unit-sets of variables.

(5) Let $F=$ the set of all unions of non-empty subsets of A_0.

(6) Let **R** = the inclusion relation restricted to F.

(7) Let **U** = the set of all items x such that $x \neq 0$ and for some variable α, $x = At(\alpha)$.

(8) Let **O** = that function which is such that

(a) the domain of **O** is the set of all sequences $\langle \delta, At(\alpha_0),\ldots, At(\alpha_{n-1})\rangle$, where δ is an **n**-place operation symbol, $\langle \alpha_0,\ldots,\alpha_{n-1}\rangle$ is an **n**-term sequence of variables, and for some variable β, $\ulcorner \delta\alpha_0\ldots\alpha_{n-1}=\beta\urcorner$ is in K^*,

(b) for each appropriate sequence in its domain, $\mathbf{O}(\langle \delta, At(\alpha_0),\ldots, At(\alpha_{n-1})\rangle)=At(\beta)$, where β is the least indexed variable such that $\ulcorner\delta\alpha_0\ldots\alpha_{n-1}=\beta\urcorner$ is in K^*.

(9) Let **P** = that function which is such that

(a) the domain of **P** is the set of all sequences $\langle \pi, At(\alpha_0),\ldots, At(\alpha_{n-2})\rangle$ where **n** is positive, π is an **n**-place predicate, $\langle\alpha_0,\ldots,\alpha_{n-2}\rangle$ is an $(\mathbf{n}-1)$-term sequence of variables, and for some variable β, $\ulcorner\pi\alpha_0\ldots\alpha_{n-2}\beta\urcorner$ is in K^*,

(b) for each appropriate sequence in its domain, $\mathbf{P}(\langle\pi, At(\alpha_0),\ldots, At(\alpha_{n-2})\rangle)=At(\gamma)$, where γ is the least indexed variable such that $\ulcorner(\forall\beta)[\beta \text{ M } \gamma \leftrightarrow \pi\alpha_0\ldots\alpha_{n-2}\beta]\urcorner$ is in K^*.

(10) Let $\mathbf{A}=$ that function whose domain is the set of all variables α such that $\ulcorner \alpha\ \mathbf{M}\ \alpha \urcorner$ is in K^*, and for each such α, $\mathbf{A}(\alpha) = At\,(\alpha)$.

(11) Let $\mathbf{M} = \langle \mathbf{R}, \mathbf{U}, \mathbf{O}, \mathbf{P}, \mathbf{A} \rangle$.

(12) $\mathbf{M}$ is a model of CB; for

(a) The part-whole relation $\mathbf{R}$ and the universe $\mathbf{U}$ are defined exactly as they were in the proof of Lemma (2.5.4.14), where it was shown that $\mathbf{U}$ is an atomistic universe of individuals for $\mathbf{R}$; and

(b) the functions $\mathbf{O}, \mathbf{P}$, and $\mathbf{A}$ satisfy the conditions (2), (3), and (4) imposed by Definition (3.5.1) on nominalistic models; for by inspection, their domains are of the right sort and their values are in $\mathbf{U}$. To show that the range of $\mathbf{P}$ is included in $\mathbf{U}$ requires an appeal to (Ax7).

(13) Show: for all formulas ϕ, $\mathbf{M}$ sat ϕ if and only if ϕ is in K^*. We proceed by simultaneous induction corresponding to the recursive characterization of terms and formulas.

(a) $\phi = \ulcorner \alpha\ \mathbf{M}\ \beta \urcorner$, for some variables α and β.

The proof of this step is exactly analogous to part (10a) in the previous proof of Lemma (2.5.4.14).

(b) $\phi = \ulcorner \alpha = \beta \urcorner$, for some variables α and β.

We use the abbreviation 'iff' for 'if and only if'.

$\mathbf{M}$ sat ϕ iff $At\,(\alpha) = At\,(\beta)$ [by Definition (4.1.3.2), Lemma (4.1.5.4), and the lines (3) and (10)], and $\ulcorner \alpha\ \mathbf{M}\ \alpha \urcorner$, $\ulcorner \beta\ \mathbf{M}\ \beta \urcorner$ are in K^*

iff $\ulcorner \alpha\ \mathbf{M}\ \alpha \urcorner$, $\ulcorner \beta\ \mathbf{M}\ \beta \urcorner$ are in K^*, and for all variables γ, if $\ulcorner At\,\gamma \urcorner$ is in K^*, then $\ulcorner \gamma \leqslant \alpha \urcorner$ is in K^* iff $\ulcorner \gamma \leqslant \beta \urcorner$ is in K^*,

iff $\ulcorner \alpha\ \mathbf{M}\ \alpha\ \& \beta\ \mathbf{M}\ \beta\ \& (\forall \gamma)\,[At\,\gamma \to (\gamma \leqslant \alpha \leftrightarrow \gamma \leqslant \beta)] \urcorner$ is in K^*,

iff $\ulcorner \alpha = \beta \urcorner$ is in K^* [by Theorem 2.5.2.19].

(c) $\phi = \ulcorner \alpha = \delta \zeta_0 \ldots \zeta_{n-1} \urcorner$, for some variable α, n-place operation symbol δ and terms $\zeta_0, \ldots \zeta_{n-1}$.

(aa) Assume that $\mathbf{M}$ sat ϕ. Then for some x, $\text{Den}\,(\mathbf{M}, \alpha, x)$ and $\text{Den}\,(\mathbf{M}, \ulcorner \delta \zeta_0 \ldots \zeta_{n-1} \urcorner, x)$. Hence, $\text{Den}\,(\mathbf{M}, \ulcorner \delta \zeta_0 \ldots \zeta_{n-1} \urcorner, At\,(\alpha))$ and $\ulcorner \alpha\ \mathbf{M}\ \alpha \urcorner$ is in K^*. Hence, for some $y_0, \ldots, y_{n-1}$, for each i, $\text{Den}\,(\mathbf{M}, \zeta_i, y_i)$ and $\langle \delta, y_0, \ldots y_{n-1} \rangle$ is in the domain of $\mathbf{O}$, and $At\,(\alpha) = \mathbf{O}(\langle \delta, y_0, \ldots, y_{n-1} \rangle)$. By (8a) for each i there is a variable β_i such that $y_i = At\,(\beta_i)$ and $\ulcorner \delta \beta_0 \ldots \beta_{n-1} = \gamma \urcorner$, for some γ, is in K^*. By (8b), $\gamma = \alpha$. By the definition of '=' and by (Ax4), $\ulcorner \beta_i\ \mathbf{M}\ \beta_i \urcorner$ is in K^* for each i, and hence $\text{Den}\,(\mathbf{M}, \beta_i, y_i)$. Thus $\mathbf{M}$ sat $\ulcorner \zeta_i = \beta_i \urcorner$. By the inductive hypothesis, $\ulcorner \zeta_i = \beta_i \urcorner$ is in K^*. By (Ax6), ϕ is in K^*.

(bb) Assume, conversely, that ϕ is in K^*. By the definition of '=',

BUNDLES OF QUALITIES, QUALITIES, AND CONCRETA   165

⌜α M α⌝ and ⌜δζ₀ ... ζ_{n-1} M δζ₀ ... ζ_{n-1}⌝ are in K^*. For each i, by (Ax4), ⌜ζ_i M ζ_i⌝ is in K^*. Clearly, for every i there is a variable $β_i$ such that ⌜ζ_i=β_i⌝ is in K^*. By the inductive hypothesis, M sat ⌜ζ_i=β_i⌝, and hence, clearly, Den(M, $ζ_i$, $At(β_i)$). By interchange of identicals, (Ax6), ⌜α=δβ₀ ... β_{n-1}⌝ is in K^*. Hence, by (8a), the sequence $\langle δ, At(β_0), ..., At(β_{n-1})\rangle$ is in the domain of O, and what O assigns to that sequence, by (8b), is $At(γ)$ for some $γ$ such that ⌜δβ₀ ... β_{n-1}=γ⌝ is in K^*. By definition, Den(M, ⌜δζ₀ ... ζ_{n-1}⌝, $At(γ)$). By the transitivity of identity, ⌜α=γ⌝ is in K^*. Hence, clearly, $At(α)=At(γ)$. Since Den(M, α, $At(α)$), M sat ϕ.

(d) $\phi=$⌜ζ=η⌝, for arbitrary terms $ζ$ and $η$.

We obtain, from (b) and (c), by induction, that all formulas of the form ⌜α=ζ⌝ satisfy the condition to be shown. (d) easily reduces to the earlier cases.

(e) $\phi=$⌜ζ M η⌝, for arbitrary terms $ζ$ and $η$.

It suffices to observe that M sat ϕ [or that ϕ is in K^*] just in case for some variables $α_0, α_1$, M satisfies ⌜ζ=α₀⌝, ⌜η=α₁⌝, and ⌜α₀ M α₁⌝ [or that these formulas are in K^*] and then to appeal to the previous inductive steps.

(f) $\phi=$⌜πζ₀ ... ζ_{n-1}⌝, for some positive integer **n**, some **n**-place predicate $π$ and terms $ζ_0, ..., ζ_{n-1}$.

(aa) Assume that M sat ϕ. Hence, by the definition of satisfaction, there exist $x_0, ..., x_{n-1}$ such that for each i, Den(M, $ζ_i, x_i$), the sequence $\langle π, x_0, ..., x_{n-2}\rangle$ is in the domain of P, and for some y in U, y is included in both x_{n-1} and in $P(\langle π, x_0, ..., x_{n-2}\rangle)$. By (9a) and since each x_i is in U, for each $i<n$ there is a variable $α_i$ such that $x_i=At(α_i)$ and ⌜α_i M α_i⌝ is in K^*. By (9b) there is a variable $γ$ such that $P(\langle π, At(α_0), ..., At(α_{n-2})\rangle)=At(γ)$ and the formula ⌜$(\forall β)[β M γ \leftrightarrow πα_0 ... α_{n-2}β]$⌝ is in K^*. For each i, Den(M, $α_i, At(α_i)$). Hence, for each i, M sat ⌜ζ_i=α_i⌝ and, by the inductive hypothesis, ⌜ζ_i=α_i⌝ is in K^*. Since x_{n-1} and $At(γ)$ have a common subset, M sat ⌜ζ_{n-1} M γ⌝. By the inductive hypothesis, ⌜ζ_{n-1} M γ⌝ is in K^*. Since, in addition, the previously mentioned generalized biconditional and also ⌜ζ_{n-1} M ζ_{n-1}⌝ are in K^*, ⌜πα₀ ... α_{n-2}ζ_{n-1}⌝ must be in K^*. By interchange of identicals, ϕ is in K^*.

(bb) Conversely, assume that ϕ is in K^*. By (Ax5), for each $i<\mathbf{n}$, ⌜ζ_i M ζ_i⌝ is in K^*. By a previous theorem and (1), for some variables $α_i$ ⌜ζ_i=α_i⌝ is in K^* and so is ⌜α_i M α_i⌝. By the inductive hypothesis,

M sat $\ulcorner\zeta_i=\alpha_i\urcorner$. Hence, Den(**M**, α_i, $At(\alpha_i)$) and Den(**M**, ζ_i, $At(\alpha_i)$). By interchange of identicals, $\ulcorner\pi\alpha_0\ldots\alpha_{n-1}\urcorner$ is in K^*. By (9a), the sequence $s=\langle\pi, At(\alpha_0),\ldots,At(\alpha_{n-2})\rangle$ is in the domain of **P**. By (9b), for some variable γ, $\mathbf{P}(s)=At(\gamma)$ and the generalized biconditional $\ulcorner(\forall\beta)\,[\beta\,\mathbf{M}\,\gamma\leftrightarrow\pi\alpha_0\ldots\alpha_{n-2}\beta]\urcorner$ is in K^*. But, $\ulcorner\alpha_{n-1}\,\mathbf{M}\,\gamma\urcorner$ is in K^*. By the inductive hypothesis, **M** sat $\ulcorner\alpha_{n-1}\,\mathbf{M}\,\gamma\urcorner$. By the definition of satisfaction, there is an x in **U**, such that x is a common subset of both $At(\alpha_{n-1})$ and of $At(\gamma)$. By inspection of the satisfaction condition (2) of (4.1.3.3), **M** sat ϕ.

(g) The sentential and universal steps in this inductive proof are treated as it is customary in completeness proofs of first-order predicate calculi which allow of denotationless terms.[2] Thus, line (13) can be regarded as shown.

We conclude that for every formula ϕ in K, **M** sat ϕ.

(4.1.5.10) THEOREM: If K is a consistent set of formulas, then there exists a model **M** such that **M** satisfies every member of K.

This theorem is derived from Lemma (4.1.5.9) in the customary manner.

(4.1.5.11) THEOREM (Strong Completeness): If K is a set of formulas and ϕ is a formula, then K (semantically) yields ϕ only if ϕ is derivable from K.

As an immediate corollary we obtain that whenever a formula ϕ is satisfied by all models of CB, then ϕ is derivable from the axioms of CB. Thus, the adequacy of CB, relative to the given interpretation, is assured.

4.2. RESEMBLANCE AND ITS CALCULUS CR

4.2.1. *Background*

In the preceding Section 4.1.3, a distinction was informally made between partial qualitative identity and partial qualitative similarity. The former notion, which we have called 'matching', was treated in the preceding section. The latter notion, which we have called 'resemblance', shall now be considered.

Informally, resemblance may be exemplified by the relation obtaining between two adjacent shades of color A and B in a color scale, if they are so arranged that A is indistinguishable by direct comparison from B, but still indirectly distinguishable by comparison with a third shade of color C which is indistinguishable by direct comparison from one of the shades A or B, but directly distinguishable from the other one.

Matching and resemblance provide for different kinds of order among bundles of qualities. In terms of matching, as we have seen, a part-whole relation among bundles of qualities can be specified: one bundle of qualities A is included in another bundle B just in case every bundle of qualities C which matches A also matches B. Resemblance cannot serve to characterize such a part-whole relation. An example will serve to show this. Suppose that

$$\{A\} \quad \{B\} \quad \{C\} \quad \{D\} \quad \ldots$$

are bundles comprising single shades of color such that all and only those bundles resemble each other which are adjacent in the given display. Suppose further than $\{A\}$ is an initial shade of color; that is, a shade which resembles only itself and $\{B\}$ (graphically, there is no other shade of color whose unit-bundle might be added on the left of the display). Then $\{B\}$ and the bundle $\{A, B\}$ resemble exactly the same bundles of colors; namely $\{A\}$, $\{B\}$, $\{C\}$, $\{A, B\}$, $\{B, C\}$, and $\{A, B, C\}$. Hence $\{B\}$ and $\{A, B\}$ are equivalent with respect to their resemblance relations. Yet $\{B\}$ is a proper part of $\{A, B\}$. Hence, one cannot generally characterize parts and wholes just in terms of resemblance. Since the resemblance relations described in this example turn out to satisfy the axioms on resemblance yet to be given, the example can be used to show that a part-whole relation cannot be defined in terms of resemblance alone within the axiomatic system to be given.

On the other hand, resemblance provides for a different kind of order which cannot be described in terms of matching, namely that of qualitative nearness. Two distinct but resembling shades of color are unrelated by partial qualitative identity in the same manner in which two completely dissimilar colors, or indeed a color and a sound, fail to have a quality in common. Unlike the matching relation, resemblance serves to determine a qualitative neighborhood of a given shade of color.

We have noted, in Section 3.3, that the notion of qualitative nearness, as well as that of sharing a qualitative part, is likely to be of interest to a nominalist for the purpose of specifying truth conditions. For this reason, we shall develop an interpreted system specifically designed to treat of the formal properties of the resemblance relation. The system is called 'CR', short for 'Calculus of Resemblance'. Apart from the notational transposition of the symbol '∘' of overlapping into the symbol 'M' for matching, which was already performed in connection with CB, the calculus CR is just an axiomatic extension of CII, to which we shall frequently refer. Intuitively, we shall continue to regard individuals as bundles of qualities. However, since we are not presently interested in schematic terms and predicates, we shall neglect the complications peculiar to the treatment of such symbols which arose in CB.

4.2.2. The Vocabulary and the Axioms of CR

The *vocabulary* of CR is just that of CII (upon replacement of the constant '∘' by 'M'), together with one further primitive constant: the two-place non-schematic predicate 'R'. Formulas of the form '$x \, R \, y$' are read 'x resembles y'.

All axioms of CII, upon uniform replacement of '∘' by 'M', are also *axioms* of CR. In addition, the following five axioms are peculiar to CR:
For all distinct variables α, β, γ, and δ,

(RAx1) $\quad \alpha \, R \, \beta \leftrightarrow (\exists \gamma)(\exists \delta) [At\gamma \, \& \, At\delta \, \& \, \gamma \leqslant \alpha \, \& \, \delta \leqslant \beta \, \& \, \gamma \, R \, \delta]$.

Informally: individuals resemble just in case each has an atomic part which resembles some atomic part of the other.

In appraising the intuitive plausibility of this axiom, it must be recalled that we continue to regard individuals as bundles of qualities. Compound bundles consisting of various qualities should resemble just in case some simple bundles typically comprising just one quality and included in the compound bundles resemble each other. For example, if we think of $\alpha = \{red_1, round_1, hard_1\}$ and $\beta = \{red_2, triangular_1, soft_1\}$, these composite bundles might resemble if the simple bundles $\{red_1\}$ and $\{red_2\}$ resemble each other.

For all variables α,

(RAx2) $\quad (\forall \alpha) \, \alpha \, R \, \alpha$.

Informally: every (existing) individual resembles itself.

For all distinct variables α, β, and γ,

(RAx3) $\quad (At\alpha \, \& \, At\beta) \rightarrow [\alpha = \beta \leftrightarrow (\forall \gamma)(\gamma \, R \, \alpha \leftrightarrow \gamma \, R \, \beta)]$.

Informally: atomic bundles of qualities are identical just in case they resemble the same individuals.

(RAx3) expresses the principle of individuation commonly held to be appropriate for bundles of qualities. There is a sense of resemblance with respect to which it is not plausible; namely that of being barely distinguishable by direct comparsion. For, imagine a world comprising just three simple qualities α, β, and γ which 'resemble' in the sense that α is just barely distinguishable from β and β from γ, while α fails to 'resemble' γ in this sense. Hence, by (RAx3), α differs from β. Yet one can bring it about, so to speak, that α becomes identical with β (according to the given axiom) by the simple expedient of annihilating the third item γ. To render (RAx3) intuitively plausible, one should translate 'x R y' by the words 'x is indistinguishable, by direct comparison, from y', while translating '$x=y$' into English by 'x is indistinguishable (by any comparison) from y'. Thus, with respect to our previous example, in the original world, α was distinguishable from β only by a comparison with γ; and α became totally indistinguishable from β once γ was removed. This accords with the intended reading of the axiom. Thus, the sense of resemblance to which we give axiomatic treatment is that of being indistinguishable by direct comparison, rather than that of being just barely distinguishable by direct comparison.

The hypothesis of (RAx3), which confines the principle of individuation to non-composite bundles of qualities, seems necessary. For, recall the example given in the previous section of the array of bundles $\{A\}$, $\{B\}$, $\{C\}$, $\{D\}$ comprising the shades of color A, B, C, and D which resemble in adjacent pairs and such that A is an initial shade of color. Under these assumptions, as we have seen, the bundles $\{B\}$ and $\{A, B\}$ resemble exactly the same things. Yet, since $\{B\}$ is a proper part of $\{A, B\}$, they should certainly not be identified. But they would be if the antecedent of (RAx3) were omitted.

For all distinct variables α, β, and γ,

(RAx4) $At\,\alpha \to (\exists \gamma)(\forall \beta)[At\,\beta \to (\beta \leqslant \gamma \leftrightarrow \beta\,\text{R}\,\alpha)]$.

Informally: If α is an atom, then there exists a bundle of qualities γ such that all and only those atoms are part of γ which resemble α. Or, the sum of all atoms resembling a given atom exists.

Recall that the calculus CII, which is our present subtheory, does not imply the existence of arbitrary sums of individuals. Hence, the existence of the sums mentioned by (RAx4) is not already assured by the underlying theory. Since individuals, in the present context, are conceived as bundles of qualities, and not necessarily as concrete things, it seems plausible to assume that all qualities which constitute the neighborhood of a given simple quality form a bundle of qualities which exists in the same universe of discourse.

For all variables α and β,

(RAx5) $\quad \alpha R \beta \to \beta R \alpha$.

This is the symmetry of resemblance. Due to (RAx1) it would have been sufficient to postulate the symmetry of resemblance just with respect to atoms.

The notion which we call 'resemblance' is approximately the same as the one which Goodman calls 'matching'.

> The difference in terminology is almost arbitrary. We feel that the word 'matching', in English, tends to connote even closer similarity than does the word 'resembling'. For this reason, we express by 'matching' the notion of partial qualitative identity, and by 'resembling' that of partial qualitative similarity. Goodman [(1966), chapters IXff.] provides a detailed discussion of the notion which we call 'resemblance' and employs it in characterizing various concepts of qualitative order. His resemblance relation, unlike ours, is restricted to atoms (p. 289, principle 9.61), it is symmetrical (principle 9.612), and reflexive when confined to atoms (principle 9.613). Our principle of individuation (RAx3) is equivalent to Goodman's principle (9.62) (given his principles 9.16, 9.612, 9.613). He introduces the notion of a 'manor', which is the individual sum of all atomic qualities resembling a given atomic quality, and expresses a counterpart of our (RAx4) by his principle (9.63) to the effect that each atomic quality has a unique manor. Goodman's informal interpretation of his system is extensive and penetrating, but no formal semantics or proofs of semantical adequacy are offered.

Subsequently, the following theorems are frequently used:

(4.2.2.1) $\quad (\alpha = \beta \,\&\, \alpha R \gamma) \to \beta R \gamma$.

(4.2.2.2) $\quad At\, \alpha \to \alpha R \alpha$.

(4.2.2.3) $\quad \alpha M \beta \to \alpha R \beta$.

The last theorem, to the effect that matching individuals resemble, follows from (Ax1), (RAx1), and (4.2.2.2).

4.2.3. *The Semantics of CR*

Informally, the individuals under consideration are still to be regarded as bundles of qualities, as discussed in the preceding sections. Accordingly, the part-whole relations which serve to interpret CR are still set-theoretic inclusion relations restricted to appropriate fields. We shall first define and then briefly explain the notion of a model for CR:

(4.2.3.1) DEFINITION: **M** is a *model* of CR if and only if there exist **R, U, A,** and **N** such that $\mathbf{M} = \langle \mathbf{R, U, A, N} \rangle$ and
(1) **R** is the inclusion relation restricted to some set,
(2) **U** is an atomistic universe of individuals for **R**,
(3) **A** is a function whose domain is a set of variables and whose range is included in **U**, and
(4) **N** is a one-one function whose domain is the set of all **R**-least elements of **U**, whose range is included in **U**, and which satisfies the following two conditions:
 (a) for every x in the domain of **N**, $x \subseteq \mathbf{N}(x)$, and
 (b) for all x and y in the domain of **N**, if $x \subseteq \mathbf{N}(y)$ then $y \subseteq \mathbf{N}(x)$.

Formally, if $\langle \mathbf{R, U, A, N} \rangle$ is a model of CR, then $\langle \mathbf{R, U, A} \rangle$ is a model of CII; and $\langle \mathbf{R, U}, 0, 0, \mathbf{A} \rangle$ is a model of CB since, in the absence of operation symbols and schematic predicates, the assignments **O** and **P** in nominalistic models may be set equal to the empty set 0.

The models of CR have the additional constituent **N**, which will be called *the neighborhood assignment* of the model. **N** is an auxiliary function which indirectly serves to interpret the resemblance relation. Informally, the neighborhood assignment is to correlate with each given simple quality (e.g., a specific shade of color) the sum of all qualities which resemble the given one. Since neighborhood assignments are one-one, it can never happen that two distinct simple qualities have the same neighborhood; that is to say, distinct simple qualities must resemble different qualities. The conditions (4a) and (4b) on models may be read 'every atom is in its own qualitative neighborhood' and 'whenever an atom x is in the neighborhood of an atom, y, then y must be in the neighborhood of x'. These conditions guarantee that the resemblance relation will be reflexive and symmetric.

For easier reference, we define fully the notion of satisfaction in models, although it deviates from Definition (2.5.3.3) only with respect to formulas containing the resemblance predicate:

(4.2.3.2) DEFINITION: **M** *sat* ϕ [in words: **M** satisfies the formula ϕ] if and only if for some **R**, **U**, **A**, and **N**, $\mathbf{M} = \langle \mathbf{R}, \mathbf{U}, \mathbf{A}, \mathbf{N} \rangle$ is a model of CR, and either:
(1) for some variables α and β, $\phi = \ulcorner \alpha \,\mathrm{M}\, \beta \urcorner$ and there exists an x in **U** such that $x \subseteq \mathbf{A}(\alpha)$ and $x \subseteq \mathbf{A}(\beta)$,
(2) for some variables α and β, $\phi = \ulcorner \alpha \,\mathrm{R}\, \beta \urcorner$ and there exist x and y in **U** such that $x \subseteq \mathbf{A}(\alpha)$, $y \subseteq \mathbf{A}(\beta)$, and $x \subseteq \mathbf{N}(y)$,
(3) for some formula ψ, $\phi = \ulcorner \sim \psi \urcorner$ and it is not the case that **M** sat ψ,
(4) for some formulas ψ and χ, $\phi = \ulcorner (\psi \to \chi) \urcorner$, and **M** sat ψ only if **M** sat χ, or
(5) for some formula ψ and variable α, $\phi = \ulcorner (\forall \alpha) \psi \urcorner$ and for all x in **U**, $\langle \mathbf{R}, \mathbf{U}, \mathbf{A}^\alpha_x, \mathbf{N} \rangle$ sat ψ.

All other semantical notions and conventions remain as previously described in Section 2.5.3.

The only novel satisfaction condition, (2), provides that two bundles of qualities resemble each other just in case each has an atomic part one of which is in the neighborhood of the other.

Again, we shall demonstrate the adequacy of our axioms relative to the given semantics: all and only those formulas are derivable from our axioms which are satisfied in all models of CR.

4.2.4. *The Semantical Adequacy of CR**

We shall frequently refer to the lemmas and theorems stated in Section 2.5.4 concerning the claculus CII, all of which continue to hold with respect to CR.

The proof that (RAx1) is valid follows easily from the satisfaction condition (2), from the properties of **N** in models, from the fact that every individual in a universe includes an atomic individual, and from Lemma (2.5.4.11).

The validity of (RAx2) follows quickly from the fact that every member

BUNDLES OF QUALITIES, QUALITIES, AND CONCRETA 173

of the universe includes an atom in that universe and from condition (4a) imposed on models.

The proof that (RAx3) is valid proceeds by showing that the neighborhoods of the two atoms in question are identical (e.g., by employing Lemma (2.3.12), the definitions, and Lemma (2.5.4.11)). It follows that the atoms themselves are identical by virtue of the fact that neighborhood assignments are one-one.

(RAx4) is shown to be valid, roughly, by assigning to the variable γ the neighborhood of the value of α, while noting that universes are closed under the neighborhoods of their atoms.

The validity of (RAx5) is obvious due to condition (4b) in the Definition (4.2.3.1) of models.

These hints suffice to suggest an easy proof of the semantical soundness of CR.

The crucial step in proving the completeness of CR is again the following Lemma:

(4.2.4.1) LEMMA: Suppose that K is a consistent set of formulas and there are infinitely many variables which are not free in any member of K. Then there exists a model **M** such that **M** satisfies every member of K.

PROOF: We begin as we did in previous proofs of this sort:

(1) By the Lemmas (2.5.4.9) and (2.5.4.8) there exists a maximally consistent extension K^* of K such that for all formulas of the form $\ulcorner(\forall\alpha)\phi\urcorner$ there is a variable β such that $\ulcorner(\phi_\beta^\alpha \vee \sim\beta \text{ M } \beta) \to (\forall\alpha)\phi\urcorner$ is in K^*.

(2) Let $At(\alpha)$ = the set of all variables β such that $\ulcorner At\,\beta\,\&\,\beta\leqslant\alpha\urcorner$ is in K^*.

(3) Let A_0 = the set of all unit-sets of variables.

(4) Let F = the set of all unions of non-empty subsets of A_0.

(5) Let **R** = the inclusion relation restricted to F.

(6) Let **U** = the set of all items x such that $x \neq 0$ and for some variable α, $x = At(\alpha)$.

(7) Let **A** = that function whose domain is the set of all variables α such that $\ulcorner\alpha \text{ M } \alpha\urcorner$ is in K^*, and for each such α, $\mathbf{A}(\alpha) = At(a)$.

(8) Let **N** = that function which is such that

(a) the domain of **N** is the set of all items $At(\alpha)$, where $\ulcorner At\,\alpha\urcorner$ is in K^*,

(b) for each $At(\alpha)$ in the domain of **N**, $\mathbf{N}(At(\alpha)) = At(\gamma)$, where γ is the least indexed variable such that the formulas $\ulcorner\gamma \text{ M } \gamma\urcorner$ and

174   NOMINALISTIC SYSTEMS

⌜$(\forall \beta) [At\ \beta \to (\beta \leqslant \gamma \leftrightarrow \beta\ R\ \alpha)]$⌝ is in K^*. The existence of such a variable is guaranteed by (RAx4).

(9) Let $\mathbf{M} = \langle \mathbf{R}, \mathbf{U}, \mathbf{A}, \mathbf{N} \rangle$.

(10) Show: $\mathbf{M}$ is a model for CR.

A comparsion with the proof of Lemma (2.5.4.14) reveals that $\langle \mathbf{R}, \mathbf{U}, \mathbf{A} \rangle$ is a model of CII. Hence, it suffices to show that $\mathbf{N}$ satisfies the conditions imposed by models of CR.

(a) The domain of $\mathbf{N}$ is the set of all $\mathbf{R}$-least elements of $\mathbf{U}$. This follows from the fact that $\langle \mathbf{R}, \mathbf{U}, \mathbf{A} \rangle$ is a model of CII and by Lemma (2.5.4.11), also by (8b) in the proof of Lemma (2.5.4.14).

(b) Clearly, the range of $\mathbf{N}$ is a subset of $\mathbf{U}$.

(c) Show: $\mathbf{N}$ is one-one.

Assume that $At\ (\alpha)$ and $At\ (\beta)$ are in the domain of $\mathbf{N}$, so that ⌜$At\ \alpha$⌝ and ⌜$At\ \beta$⌝ are in K^*, and that $\mathbf{N}(At\ (\alpha)) = \mathbf{N}(At\ (\beta))$. For some γ and γ', $\mathbf{N}(At\ (\alpha)) = At\ (\gamma)$ and ⌜$\gamma\ M\ \gamma$⌝, ⌜$(\forall \beta) [At\ \beta \to (\beta \leqslant \gamma \leftrightarrow \beta\ R\ \alpha)]$⌝ are in K^*, and $\mathbf{N}(At\ (\beta)) = At\ (\gamma')$ and ⌜$\gamma'\ M\ \gamma'$⌝, ⌜$(\forall \beta') [At\ \beta' \to (\beta' \leqslant \gamma' \leftrightarrow \beta'\ R\ \beta)]$⌝ is in K^*.

(aa) Show: ⌜$\delta\ R\ \alpha$⌝ is in K^* just in case ⌜$\delta\ R\ \beta$⌝ is in K^*, ($\delta \neq \alpha, \beta$). From left to right, assume that ⌜$\delta\ R\ \alpha$⌝ is in K^*. Since ⌜$At\ \alpha$⌝ is in K^* and by (RAx1), for some ε, ⌜$At\ \varepsilon\ \&\ \varepsilon \leqslant \delta\ \&\ \varepsilon\ R\ \alpha$⌝ is in K^*. Hence, ⌜$\varepsilon \leqslant \gamma$⌝ is in K^*. Since $At\ (\gamma) = At\ (\gamma')$ and ε is in $At\ (\gamma)$, ⌜$\varepsilon \leqslant \gamma'$⌝ is in K^*, and hence so is ⌜$\varepsilon\ R\ \beta$⌝. Since ⌜$At\ \beta$⌝ and clearly also ⌜$\beta \leqslant \beta$⌝ are in K^* and by (RAx1), ⌜$\delta\ R\ \beta$⌝ is in K^*. The implication from right to left is similar, establishing (aa).

Hence, by (1) and (aa), ⌜$(\forall \delta)(\delta\ R\ \alpha \leftrightarrow \delta\ R\ \beta)$⌝ is in K^*. By (RAx3), ⌜$\alpha = \beta$⌝ is in K^*. Hence, clearly, $At\ (\alpha) = At\ (\beta)$, establishing (c).

(d) Show: for all $At(\alpha)$ and $At(\beta)$ in the domain of $\mathbf{N}$, if $At(\alpha) \subseteq \mathbf{N}(At(\beta))$ then $At\ (\beta) \subseteq \mathbf{N}(At\ (\alpha))$.

Assume the hypothesis, so that ⌜$At\ \alpha$⌝, ⌜$At\ \beta$⌝ are in K^*. For some γ, γ', $\mathbf{N}(At\ (\alpha)) = At\ (\gamma)$ and ⌜$\gamma\ M\ \gamma$⌝, ⌜$(\forall \beta) [At\ \beta \to (\beta \leqslant \gamma \leftrightarrow \beta\ R\ \alpha)]$⌝ are in K^*, $\mathbf{N}(At\ (\beta)) = At\ (\gamma')$ and ⌜$\gamma'\ M\ \gamma'$⌝, ⌜$(\forall \beta') [At\ \beta' \to (\beta' \leqslant \gamma' \leftrightarrow \beta'\ R\ \beta)]$⌝ is in K^*. To show that $At\ (\beta) \subseteq \mathbf{N}(At\ (\alpha))$, assume that δ is in $At\ (\beta)$, so that ⌜$At\ \delta\ \&\ \delta \leqslant \beta$⌝ is in K^*. Since $At\ (\alpha) \subseteq \mathbf{N}(At\ (\beta)) = At(\gamma')$, and α is in $At\ (\alpha)$, clearly ⌜$\alpha \leqslant \gamma'$⌝ is in K^*. Hence, ⌜$\alpha\ R\ \beta$⌝ is in K^*. By (RAx5), ⌜$\beta\ R\ \alpha$⌝ is in K^*, and hence ⌜$\beta \leqslant \gamma$⌝ is in K^*. By transitivity, ⌜$\delta \leqslant \gamma$⌝ is in K^*. Therefore, δ is in $\mathbf{N}(At\ (\alpha))$, which completes the proof of (d).

(e) Show: for all $At\ (\alpha)$ in the domain of $\mathbf{N}$, $At\ (\alpha) \subseteq \mathbf{N}(At\ (\alpha))$.

The proof of this step is roughly similar but easier than the preceding one.

We conclude that **M** is indeed a model for CR.

(11) Show: for all formulas ϕ, **M** sat ϕ if and only if ϕ is in K^*. The proof proceeds by induction on ϕ.

(a) $\phi = \ulcorner \alpha \, R \, \beta \urcorner$, for some variables α and β.

One shows, by a subsidiary proof, that for all α and β if $At(\alpha)$ and $At(\beta)$ are in **U**, then $At(\alpha) \subseteq At(\beta)$ whenever $\ulcorner \alpha \leqslant \beta \urcorner$ is in K^*. Then there is no difficulty in proving this step by using (RAx1). All other steps in the induction remain exactly as in the proof of Lemma (2.5.4.14).

These considerations suffice to demonstrate the strong semantical completeness of CR.

4.3. QUALITIES AND THEIR CALCULUS CQ

4.3.1. Background

Previously, three nominalistic positions have been distinguished with reference to the increasingly stringent demands that theories should refer only to individuals, that they should only refer to individuals which are non-sets, or that they should only refer to individuals which are concrete. In Section 4.1 we have provided a formal system so interpreted that its expressions refer to bundles of qualities, that is, to certain individuals which also happen to be sets. We shall now try to meet the demands of those nominalists who will not be committed to sets of any sort, although they are willing to endorse other kinds of abstract individuals.

The abstract individuals of which we shall treat are thought of as simple and compound qualities. The notion of a quality, in this connection, should not be identified with that of a property if the word 'property' is taken to refer to arbitrary intensions expressed by predicates or formulas. While properties are part of the subject matter of intensional logic[3], our theory of qualities will remain quite extensional and is not designed to account for the difficulties of oblique contexts. Instead, we regard qualities, roughly speaking, as those positive, natural, and striking features with respect to which it is customary to say that concrete things resemble one another. Rather than to define the notion of a quality, we shall characterize it indirectly by showing, both informally and formally, how statements regarding qualities have counterparts among the statements which concern bundles of qualities. We shall first illustrate informally how discourse concerning bundles of qualities can be translated into discourse concerning qualities themselves.

The pre-systematic connotations of the words 'matching', 'resembling', 'quality', 'bundle of qualities' seem such that the following statements appear true:

(4.3.1.1) A quality A is 'simple' (that is, has no proper qualitative parts) just in case the bundle which comprises all qualitative parts of A is 'atomic' (that is, has no bundles of qualities for a proper part).

Thus, the quality 'yellow' is regarded as a simple quality just in case the bundle comprising 'yellow' alone is regarded as an atomic bundle of qualities. If 'yellow' is not regarded as simple, but as having, e.g., the shade 'ocher' for a proper qualitative part, then the bundle comprising just the quality 'ocher' is regarded as properly included in the bundle comprising all shades of color within the range of colors to which the predicate 'is yellow' properly applies.

(4.3.1.2) A quality A is 'simple' just in case it is 'uniform'; and A is 'uniform' just in case any two qualities which match A match one another.

(4.3.1.3) A bundle of qualities is 'atomic' just in case any two bundles which match it match one another. (upon replacement of '$\circ$' by 'M', this is just Theorem (2.5.2.36) of CII).

For example, if a concrete object is not uniformly colored, then it must exhibit at least two distinct shades of color A and B, which makes it possible for two objects of color A and B respectively to match the given one without matching one another. It must be recalled that our notion of matching is that of partial qualitative identity, not that of similarity. These informal considerations suggest that statements regarding 'simple' qualities have counterparts among the statements concerning 'atomic' bundles of qualities. Just as the notion of an 'atom' is relative to a universe of discourse, so also qualities should be 'simple' only relative to a given universe of qualities.

Further,

(4.3.1.4) If A and B are qualities and if every quality which matches A also matches B, then A is a qualitative part of B.

(4.3.1.5) If A and B are bundles of qualities and if every bundle which matches A also matches B, then A is included in B.

Thus, the quality 'ocher' is part of the compound quality 'round-and-ocher' just in case every quality which matches 'ocher' also matches 'round-and-ocher'.

These informal comparisons suggest that the part-whole relation which obtains between simple and compound qualities has the same structure as the inclusion relation which holds between smaller and larger bundles of qualities. This isomorphism, made formally explicit, will serve to provide a new interpretation for the formal system CB in such a way that the former reference to sets (bundles of qualities) is replaced by reference to non-sets (qualities). The resulting system CQ (short for 'calculus of qualities') will differ from the former system CB only with respect to its semantics. In constructing CQ, we shall take the following steps:

(1) The set theory without individuals which is presently expressed in our meta-language will be replaced by a set theory whose variables range over both sets and non-sets.

(2) Additional axioms are introduced in the meta-theory to ensure that there exists at least one part-whole relation among non-sets of the appropriate sorts.

(3) Certain isomorphisms are exhibited between the part-whole relations which are confined to non-sets and those which served to interpret CB.

(4) A new interpretation is provided for the axioms of CB so that, roughly speaking, reference to bundles of qualities is replaced by reference to qualities. The resulting interpreted system is called 'CQ'.

(5) The semantical interpretations of CB and of CQ are shown to be equivalent in the appropriate sense.

(6) From (5) and from the semantical adequacy of CB we shall be able to infer the semantical adequacy of CQ.

4.3.2. *The Meta-Theory of CQ*

The meta-language of CQ is to express a set theory with 'individuals' in the sense of non-sets. Up to now there would have been little justification in adopting a set theory with individuals, since we were not prepared to specify any positive axioms concerning them. Indeed, the two senses of the word 'individual' (as 'member of a universe of individuals', and as 'non-set') might have led to misunderstandings. But presently we shall need a set theory with non-sets. To be definite, the axioms of the meta-theory shall include all those formulated by Suppes in *Axiomatic Set Theory* (1960), with some notational changes to conform to the set-theoretic notation so far employed. The one-place predicate 'Σ' of the meta-language will be read 'is a set'. All Capital Latin letters other than 'M', all Greek letters and all script letters shall refer to sets exclusively. Lower case Latin letters refer to either sets or non-sets. As before, new expressions will be introduced into the meta-language whenever doing so is convenient.

If it were our objective merely to secure a part-whole relation among non-sets, it would suffice to add an axiom to the effect that the set of non-sets has the power of the continuum. A part-whole relation on non-sets could then be induced as any relation, among non-sets, which is isomorphic to a part-whole relation, as earlier defined, among sets. Instead, we shall postulate the existence of a special part-whole relation among non-sets. In doing so we shall secure at least one part-whole relation among non-sets even if we should subsequently decide to weaken or to eliminate altogether that portion of the meta-theory which deals specifically with sets. Furthermore, we would like to assert, concerning non-sets, that they are actually individuals in the sense, e.g., of satisfying Goodman's principle of individuation with respect to a given part-whole relation whose specification does not require an appeal to the existence of certain sets. Furthermore, we would like to assert in the meta-language, concerning non-sets, that they are actually qualities. This could be achieved by introducing the predicate 'is a quality' and by presupposing its customary meaning. However, this predicate would not give rise to a part-whole relation. Instead, we shall introduce as a primitive the two-place predicate 'M' for 'matches' whose systematic role is known to us from the calculus CB. The relation of matching may obtain between sets (e.g., between bundles of qualities), and need not be defined on all non-sets. However, we shall only be interested in the matching relation in so far as it applies to non-sets; and we shall intuitively regard non-sets which match themselves as qualities.

BUNDLES OF QUALITIES, QUALITIES, AND CONCRETA 179

The following three definitions are counterparts, in the meta-language, of the definitions (D1)–(D3) stated in Section 4.1.4. In order to avoid conflicting notation in the meta-language, the symbols '$\leqslant$' and '$<$' of CB are here replaced by the symbols '$\leqslant *$' and '$<*$':

(Def. 1) $x \leqslant * y$ if and only if x M x and for every z, if z M x then z M y.
(Def. 2) $x <* y$ if and only if $x \leqslant * y$ and not $y \leqslant * x$.
(Def. 3) Atx if and only if x M x and there is no y such that $y <* x$.

Beyond the customary axioms of set theory, the following additional *proper axioms* determine the needed formal properties of non-sets:

(A1)   There are infinitely many atomic non-sets.

If it seems undesirable that (A1), under its present formulation, presupposes the existence of functions, it might be replaced by an axiom-schema comprehending, for every natural number **n**, the assertion that there exist at least **n** atomic non-sets.

(A2)   If $x \leqslant * y$ and $y \leqslant * x$, then $x = y$.

In words: items which are part of one another are identical. This axiom constitutes a principle of individuation for qualities (that is, for non-sets which match themselves).

(A3)   If $\sim \Sigma x$ and $\sim \Sigma y$, then x M y just in case for some z, Atz and $z \leqslant * x$ and $z \leqslant * y$.

In words: non-sets match just in case they have a common atomic part. This axiom is the counterpart of (Ax3) in CB.

We adopt as an axiom-scheme the sum-axiom (AxS6) of CIIII. That is, if 'y' does not occur in the formula ϕ,

(AS4)   If there exists an x such that $\sim \Sigma x$ and Atx and ϕ, then there exists a y such that $\sim \Sigma y$ and for every x, if $\sim \Sigma x$ and Atx then $x \leqslant * y$ just in case that ϕ.

Informally: If some atomic non-set satisfies the condition ϕ, then there exists a certain non-set y such that exactly those atomic non-sets are part of y which satisfy the condition ϕ. Or: If some atomic non-set satisfies the condition ϕ, then the sum exists of all atoms which satisfy ϕ, and this sum is a non-set.

(A5) $\sim 0 \text{ M } 0$.

In words: the empty set does not match itself. Since we have earlier agreed to set all improper descriptions of the meta-language equal to the empty set 0, this axiom has the effect that every descriptive phrase which refers to an item which matches itself will be a proper description. In addition, it seems to accord with the spirit of nominalism not to construe the empty set (e.g., the empty bundle of qualities) as a true self-matching individual.

(A6) If $\sim \Sigma x$ and either $y \text{ M } x$ or $x \text{ M } y$, then $\sim \Sigma y$.

In words: everything which matches a non-set is itself a non-set.

We state, without proof, a number of lemmas:

(4.3.2.1) There exist infinitely many non-sets which match themselves.

This follows from (A1), (A6), and the definitions.

(4.3.2.2) There exist infinitely many non-sets which are atoms.

This follows from (4.3.2.1), (A6), (A3), if (A1) is replaced by (4.3.2.1).

(4.3.2.3) If A is a non-empty set of atomic non-sets, then there exists a non-set y such that exactly those atomic non-sets are part of y which are members of A.

This follows from (AS4) by replacing 'ϕ' by 'x is a member of A'.

(Def. 4) the *sum* of all things in A = the unique object y such that $\sim \Sigma y$, A is a non-empty set of atomic non-sets, and exactly those atomic non-sets are part of y which are members of A.

If one of the conditions in the definiens is not satisfied, then the sum of all things in A will be the empty set.

(4.3.2.4) If x is an atomic non-set and part of the sum of all things in A, then x is a member of A.

This lemma will be useful in the consideration of part-whole relations. Its proof appeals to (Def. 4) and (A6).

BUNDLES OF QUALITIES, QUALITIES, AND CONCRETA 181

(4.3.2.5) Let $y=$ the sum of all things in S and assume that y M y. then (1) for every x in S, $x \leqslant *y$, and (2) for all z, assuming that for every x in S $x \leqslant *z$, $y \leqslant *z$.

Thus, the sum of all items in S, if it exists, is just the least upper bound or supremum, with respect to '$\leqslant *$', of the things in S. The proof appeals to (A5), (A6), (A3), and the definitions.

(4.3.2.6) Suppose that $\sim \Sigma x$ and x M x. Then Atx if and only if for every y and z, if y M x and z M x then y M z.

In words: a self-matching non-set is atomic just in case any two items which match it match one another. This lemma corresponds to our intuition, expressed earlier, that all and only those qualities are simple which are uniform.

In addition to the axioms so far mentioned, we shall presuppose, as part of a set theory with individuals, that the set of all non-sets exists.

We shall presently employ this information in the meta-language concerning non-sets to characterize part-whole relations and new models appropriate to the axioms of CB whose universes comprise non-sets. Both intuitively and formally these developments will not hold any great surprises and may be omitted without loss of continuity.

4.3.3. *The Interpreted System CQ**

Both the *vocabulary* and the *axioms* of CQ are those of CB.

Confirming to the program outlined previously, the fields of part-whole relations appropriate to CQ shall consist only of self-matching non-sets which we regard as qualities:

(4.3.3.1) DEFINITION: **R** is a *qualitative part-whole relation* just in case for some infinite set A of atomic non-sets, **R**=the set of all pairs (x, y) such that both x and y are sums of non-empty subsets of A and $x \leqslant *y$.

Due to (A5) and the Lemmas (4.3.2.2) and (4.3.2.4), qualitative part-whole relations are indeed part-whole relations as previously defined by (2.2.6), and there exists at least one qualitative part-whole relation.

The *models* of CQ are nominalistic models, as defined by (3.5.1), whose first constituent **R** is a qualitative part-whole relation. By the *universe of* a model is meant the second constituent **U** of the model. The notions of *denotation* ('Den(**M**, τ, x)'), of *satisfaction* ('**M** sat φ'), and of *truth* relative to a model **M** remain as previously specified by the respective definitions (4.1.3.2), (4.1.3.3), and (4.1.3.4), except that all reference to the models of CB is replaced by reference to the models of CQ.

The extra-systematic interpretation carried to this formalism is briefly the following: The items which enter a qualitative part-whole relation are regarded as possible qualities, both simple and complex. Given a universe of individuals relative to such a part-whole relation, its members are thought of as those qualities which happen to be 'actualized' in that universe. The atoms within a given universe are taken to be the simplest qualities among those which happen to occur in that universe. Thus, a particular universe might comprise all shades of color which can be discriminated by a given person. If so, an atom in that universe can be regarded as a least discriminable shade of color discernible by that person. The truth conditions, loosely paraphrased, provide that two qualities in a given universe match just in case they have a common qualitative part in that universe. Each one-place predicate designates a quality in the universe. Thus, the predicate 'is ocher' might designate a quality which is simple or uniform relative to the given universe, while the predicate 'is yellow' might designate a compound quality of which the designatum of 'is ocher' is a proper qualitative part. Things are conceived as compound qualities (as sums, rather than as bundles of qualities) whose simplest qualitative parts are just those specific qualities which can be truly ascribed to them. A one-place predicate is applicable to an object just in case its designatum matches that object. For further informal discussions, we refer to Sections 4.1.1 and 4.1.2, which can easily be made relevant to the present topic just by translating remarks concerning bundles of qualities into ones concerning qualities themselves.

There will be occasion to speak of isomorphic nominalistic models. In defining this notion, we shall presuppose an understanding of isomorphic relations (see Definition 1.3.1.53):

(4.3.3.2) DEFINITION: Suppose that $\langle V, L \rangle$ is a formal system and that $\mathbf{M} = \langle \mathbf{R}, \mathbf{U}, \mathbf{O}, \mathbf{P}, \mathbf{A} \rangle$ and $\mathbf{M}' = \langle \mathbf{R}', \mathbf{U}', \mathbf{O}', \mathbf{P}', \mathbf{A}' \rangle$ are nominalistic models for $\langle V, L \rangle$. Then **M** is *isomorphic, by f, with* **M**' just in case the following conditions are satisfied:

(1) **R** is isomorphic, by f, with **R′**,
(2) the restriction of **R** to **U** is isomorphic, by f, with the restriction of **R′** to **U′**,
(3) for every natural number **n** and for every **n**-place operation symbol δ in V, $\langle \delta, x_0, \ldots, x_{n-1} \rangle$ is in the domain of **O** if and only if $\langle \delta, f(x_0), \ldots, f(x_{n-1}) \rangle$ is in the domain of **O′**, and for every sequence $\langle \delta, x_0, \ldots, x_{n-1} \rangle$ in the domain of **O**, $f(\mathbf{O}(\langle \delta, x_0, \ldots, x_{n-1} \rangle)) = \mathbf{O}'(\langle \delta, f(x_0), \ldots, f(x_{n-1}) \rangle)$,
(4) for every natural number **n** and for every $(n+1)$-place predicate π in V, $\langle \pi, x_0, \ldots, x_{n-1} \rangle$ is in the domain of **P** if and only if $\langle \pi, f(x_0), \ldots, f(x_{n-1}) \rangle$ is in the domain of **P′**, and for every sequence $\langle \pi, x_0, \ldots, x_{n-1} \rangle$ in the domain of **P**, $f(\mathbf{P})\langle \pi, x_0, \ldots, x_{n-1} \rangle) = \mathbf{P}'(\langle \pi, f(x_0), \ldots, f(x_{n-1}) \rangle)$,
(5) the domain of **A** = the domain of **A′**, and for every variable α in the domain of **A**, $f(\mathbf{A}(\alpha)) = \mathbf{A}'(\alpha)$.

Further, **M** is said to be *isomorphic* with **M′** just in case for some f, **M** is isomorphic, by f, with **M′**.

This definition of isomorphism is quite as one would expect it to be. The next theorem guarantees that one can always construct a model for CQ on the basis of any denumerable model of CB:

(4.3.3.3) THEOREM: Suppose that $\mathbf{M} = \langle \mathbf{R}, \mathbf{U}, \mathbf{O}, \mathbf{P}, \mathbf{A} \rangle$ is a model of CB and that the set of all **R**-least elements is denumerable. Then there exists a model **M′** for CQ such that **M** is isomorphic with **M′**.

The proof, in outline, proceeds as follows: We know, by Lemma (4.3.2.2), that there exist infinitely many non-sets which are atoms. Let g be a one-one mapping of the **R**-least elements onto a denumerable set of atomic non-sets. Let f be a function whose domain is the field of **R** and which assigns to every element x in that field the sum of all items $g(y)$, where y is **R**-least and bears **R** to x. Let **R′** be the relation $\leq^*$ restricted to the range of f. It is not hard to show, with special reference to (4.3.2.4) and (4.3.2.5), that for every x and y in the field of **R**, x **R** y if and only if $f(x) \leq^* f(y)$. From this it follows immediately that f is one-one, and that **R** is isomorphic, by f, with **R′**. Let **U′** be the image, under f, of **U**. Clearly, the restriction of **R** to **U** is isomorphic, by f, to the restriction of **R′** to **U′**. Let **O′** and **P′** satisfy the respective conditions (3) and (4) of the Definition (4.3.3.2). Also, let **A′** satisfy the condition (5) of that definition. Clearly, **M** is isomorphic, by f, with the model $\langle \mathbf{R}', \mathbf{U}', \mathbf{O}', \mathbf{P}', \mathbf{A}' \rangle$ of CQ.

Conversely, every model for CQ gives rise to an isomorphic model for CB:

(4.3.3.4) THEOREM: Suppose that $\mathbf{M} = \langle \mathbf{R, U, O, P, A} \rangle$ is a model for CQ. Then there exists a model $\mathbf{M}'$ for CB such that $\mathbf{M}$ is isomorphic with $\mathbf{M}'$.

The essentials of the proof are these: By some one-one function g map the atomic non-sets in the field of $\mathbf{R}$ onto their own singletons. Let $F =$ the set of all non-empty unions of such singletons. Let $\mathbf{R}' =$ the inclusion relation restricted to F. Then $\mathbf{R}'$ will be a part-whole relation appropriate to models of CB. Let f be that function whose domain is the field of $\mathbf{R}$ and which assigns to every element x in that field the union of all items $g(y)$, where y is an atomic non-set in the field of $\mathbf{R}$ and $y \leqslant {}^* x$. One shows that for every x and y in the field of $\mathbf{R}$, $x \leqslant {}^* y$ if and only if $f(x) \subseteq f(y)$. Clearly, f is one-one and $\mathbf{R}$ is isomorphic, by f, with $\mathbf{R}'$. Let $\mathbf{U}'$ be the image, under f, of $\mathbf{U}$. Let $\mathbf{O}'$, $\mathbf{P}'$, and $\mathbf{A}'$ satisfy the conditions (3), (4), and (5) respectively of Definition (4.3.3.2). Then $\langle \mathbf{R}', \mathbf{U}', \mathbf{O}', \mathbf{P}', \mathbf{A}' \rangle$ is a model of CB.

(4.3.3.5) THEOREM: Suppose that $\mathbf{M}$ is a model for CB, $\mathbf{M}'$ is a model for CQ, and $\mathbf{M}$ is isomorphic with $\mathbf{M}'$. Then, for all formulas ϕ, $\mathbf{M}$ sat ϕ (in CB) if and only if $\mathbf{M}'$ sat ϕ (in CQ).

One shows first, by an induction on terms τ, that for every x, $\text{Den}(\mathbf{M}, \tau, x)$ if and only if $\text{Den}(\mathbf{M}', \tau, f(x))$, where f is the isomorphism from $\mathbf{M}$ to $\mathbf{M}'$. Then one shows the conclusion of the theorem by an induction on ϕ.

It is now an easy matter to demonstrate the semantical adequacy of CQ. Its soundness follows from the soundness of CB, together with Theorems (4.3.3.4) and (4.3.3.5). The semantical completeness is a consequence of the completeness of CB, the observation that the set of atoms in the part-whole relation $\mathbf{R}$ in the model used to prove this completeness is denumerable, and by appealing to Theorems (4.3.3.3) and (4.3.3.5).

4.4. CONCRETA

The interpreted formal system CB will satisfy, as has been observed, the requirements of a liberal nominalist who is willing to countenance any individuals whatever, even individuals which are sets. Criteria other than the barest demands of nominalism may require that the universes of individuals which serve to interpret a system shall comprise only non-sets. Such additional criteria are easily met by the system CQ presented in the preceding section. However, the subject matter of that system, namely qualities, still concerns abstract individuals. We shall now turn to the stringent requirement, not met by any system so far considered, that all individuals shall be concrete. To begin with, at least an informal explanation may be helpful, of how the word 'concrete' will be used.

In philosophical discourse, the expressions 'physical', 'material', 'particular', 'specific', and 'individual' all seem to differ, in meaning, from the word 'concrete'. Physical objects, presumably the entities investigated in physics, include theoretical items, such as magnetic fields, which one would not call 'concrete'. Material objects, which we take to be items characterized by position, mass, and the like, include particles which are not accessible to the senses and to that extent not concrete. Among particulars, one might classify such non-concrete entities as numbers; and among specific things one lists recurrent shades of color. And individuals, as items which are individuated by certain principles of which Goodman's is an example, need not be concrete in any respect. Of course, these concepts are frequently ill distinguished or distinguished with respect to different principles in other contexts.

A paradigm example of a concrete thing, it would seem, is an object which can be handled and kicked, measured and compared, looked at and photographed from all sides, and preferably heard, smelled and tasted as well. Concreta may be construed as bundles of qualities. But not every bundle of qualities is a concretum. For example, a bundle comprising just one simple quality, or one comprising just a range of pairwise resembling qualities would not be regarded as a concretum.

Goodman (1966) proposed an analysis of 'concretum' which relies intuitively on position qualia and formally on the special primitive 'togetherness' introduced for that very purpose. According to that account, an example of a concretum (of a minimally concrete thing) might be a color-spot-moment. The fact that the color of that item occurs 'in' the given spot 'at' the given moment while the item is not further qualified is expressed by saying that the color is 'together' with the spot, the moment, and their sum, that the spot is 'together' with the color, the moment, and their sum, that the moment is 'together' with the color, the spot and their sum, while there is no other quality (say, no sound and no hardness) which might in addition be 'together' with the items already mentioned. Goodman (1966) makes only extra-systematic use of position qualia: they are not formally distinguishable from other qualia, and they do not serve formally in establishing positional order.

Furthermore, for reasons which were mentioned in Section 2.10.3, we doubt whether the introduction of position qualia would serve in characterizing sequential individuals. Since their systematic import seems just as unclear as their epistemological primacy, we would prefer to leave position qualia out of account. But once we do, Goodman's notion of a concretum becomes unintuitive. Indeed, even if one grants positional qualities and an understanding of 'togetherness', it seems somewhat contrary to customary usage to call an item 'concrete' which need consist of no more than one specific shade of color at a given time and place.

We shall not attempt to define the predicate 'concrete'. Instead we draw attention to a feature of certain formal systems which will be called 'concretistic' systems as opposed to 'abstractistic' systems. Attribution of these dreadful but conventient labels is connected with the difficulty which Goodman calls 'the problem of abstraction'[4], and which consists, roughly speaking, in classifying individuals with respect to their qualities, given that we permit no reference to items which are not concrete.

The problem of abstraction arises for us in the following form: Given a language with schematic predicates and a universe of discourse, we have decided (for reasons detailed in Section 3.3) that each predicate shall designate some item in that universe. If the universe of discourse comprises only concrete individuals, then each predicate should designate some concrete thing. Thus, the predicate 'is white' might designate a concrete exemplar of the color 'white'. We shall further need a relation R of predication satisfying the condition that

(4.4.1)   a sentence of the form 'A is white' is true just in case the item named by 'A' bears R to the exemplar assigned to the predicate 'is white'.

As long as abstract individuals are admitted (such as the color 'white', or a range of colors which fall under the appellation 'white'), the relation R of predication can be specified in terms of the part-whole relation. Thus, in the calculus CB, the relation of predication was that of matching; that is, that of having a common (qualitative) part (see Section 4.1.3). Hence, in CB, the relation of predication was definable in terms of that part-whole relation which is appropriate to bundles of qualities. If abstract individuals of a different sort were admitted, such as the sum of all white things throughout space and time, then the relation of predication could again be specified in terms of the part-whole relation; for 'to be white', in such a system, could mean to be part of that sum. It is clear, however, that no plausible relation of predication can be defined in terms of that part-whole relation into which only those individuals enter that would normally be called 'concrete'. That is to say, there is no formula $\phi(x, y)$ of some pure calculus of individuals treating only of the part-whole relation (say of CI I) of such a sort that for every universe U comprising only concrete individuals, if y is the exemplar of the property 'white', then the respective objects x and y satisfy the formula $\phi(x, y)$ exactly when x is white. To see this intuitively, it suffices to consider a universe U comprising just two concrete atoms x and y, where y exemplifies the color 'white', while it is undetermined whether or not x is white. All relations $\phi(x, y)$ which are specifiable just in terms of the part-whole relation confined to U, and which are true of x and y, will simply be coextensional with '$x \neq y$'; and that formula does not determine whether or not x is white.

Intuitively, qualities may indeed be 'parts' of concrete things, but not in the same sense of 'part' in which concrete things have other concrete things for 'parts'. The 'mode of composition' with respect to which a typewriter is composed of various mechanical parts is not the same as that with respect to which it is composed of whatever noticeable qualities it has, even if we admit, in different senses of the word 'whole', that the typewriter is nothing but a whole of its mechanical parts, and nothing but a whole composed of its qualities.

In so far as it appears in the interpretation of sentences having subject-predicate form, the problem of abstraction seems to consist in the problem of specifying a relation of predication which is not definable in terms of the same part-whole relation with respect to which the chosen individuals are parts or wholes. Note that in formulating the problem in this fashion, there was no need to employ the expressions 'concrete' or 'abstract'. It seems natural, therefore, to define a nominalistic system with schematic predicates as *abstractistic* just in case its relation of predication is definable in terms of its part-whole relation; and to call it *concretistic* otherwise. By this criterion, both of the calculi CB and CQ are abstractistic, since the relation of predication in both systems is that of having a common (qualitative) part.

Consider nominalistic systems whose object-language comprises no primitive symbols other than those of the pure calculus of individuals and schematic predicates. In abstractistic systems of this sort, but not in concretistic ones, it is possible to express in the object-language a counterpart of the condition which in the meta-language serves to define truth for sentences of subject-predicate form. Thus, in the calculus CB, the axiom (Ax7) can be taken to assert that whenever a predicate applies at all, then exactly those sentences of subject-predicate form are true in which the subject matches the designatum of the predicate. This possibility of expressing syntactical counterparts of semantical conditions was used in achieving the completeness of abstractistic systems. Since concretistic systems lack this feature, they will pose, as one of the 'problems of abstraction', the difficulty of rendering such systems semantically complete.

For reasons which were more fully discussed in Sections 3.3 and 3.4, we select as the relation of predication some relation of similarity. Thus, all and only those concrete things may be called 'white' which resemble the concrete exemplar of the color 'white'. However, concrete things invariably possess more than one quality with respect to which they can resemble other things. Thus, a truth condition like the following would be clearly inadequate:

(4.4.2)   a sentence of the form 'A is white' is true just in case the item named by 'A' resembles the exemplar designated by 'is white'.

The concrete exemplar of 'white' might, e.g., be soft. If so, according to (4.4.2), the predicate 'is white' would turn out to be applicable to things which are soft though not white.

One might consider assigning to the predicate 'is white' two or more exemplars of sufficient variety so that no quality other than that of being white is common to them all (e.g., one of them might be white and soft, and another white and non-soft). Then one might specify that exactly those individuals shall be called 'white' which resemble all of the exemplars. However, it seems implausible that every possible universe of concreta should comprise individuals of the required variety.[5]

It seems more promising to employ, as a relation of predication, the relation of *resembling in a given respect*. Thus, concreta might be said to resemble with respect to color, with respect to shape, and so forth.

In order to characterize this concept, we adopt as our meta-theory that set-theory with individuals which was previously specified in Section 4.3.2. In addition, we impose in the meta-language on all self-matching non-sets the axioms (RAx1)–(RAx5) concerning the notion of resemblance (see Section 4.2.2).

Two auxiliary notions are introduced in the meta-language: that of a 'clan' and that of a 'category'. Both notions are due to Goodman.[6] Informally, a clan is to be a sum of simple qualities satisfying the condition that any two simple qualities in the sum are connected by a chain of resembling qualities in that sum. Thus, an unbroken segment of a color scale, or of a scale of sounds arranged by pitch, is an intuitive example of a clan. But a whole composed of both sounds and colors is not regarded as a clan, since there is no unbroken chain of resembling qualities between any one color and any one sound. A clan is defined by the condition that if it is broken up into any two pieces x and y then some simple quality in x must resemble some simple quality in y:

(4.4.3)   DEFINITION: x is a *clan* just in case $\sim \Sigma x$ and x M x and for every y: if $y <^* x$ then there exist u and v such that $\sim \Sigma u$ and $\sim \Sigma y$ and Atu and Atv and $u \leqslant^* y$ and $(v \leqslant^* x$ and $\sim v \leqslant^* y)$ and u R v.

Informally: x is a clan just in case x is a self-matching non-set (intuitively: a sum of qualities) and for every proper part y of x there exist simple qualities u and v such that u is part of y, v is part of x but not of y and u resembles v. Still more loosely: x is a clan if any two discrete parts of x, namely y and the difference between x and y, partially resemble one another.

Examples of 'categories', intuitively, are the sum of all colors; also the sum of all sounds; generally, the sum of all qualities which resemble something or other in a given respect.

(4.4.4)   DEFINITION: x is a *category* just in case x is a clan and, for every y, if y is a clan then $\sim x <^* y$.

Informally: a category is a clan which is not a proper part of any clan. Or: a category is a most comprehensive clan.

One can now readily define the notions of matching and of resembling in a given respect, the respect in question being a category:

(4.4.5)   DEFINITION: *x matches y with respect to c* just in case c is a category and for some z, $z \leqslant^* c$, $z \leqslant^* x$, and $z \leqslant^* y$.

(4.4.6)   DEFINITION: *x resembles y with respect to c* just in case c is a category and for some u and v, $u \leqslant^* c$, $v \leqslant^* c$, $u \leqslant^* x$, $v \leqslant^* y$, and $u \, R \, v$.

Informally: two items match with respect to a category just in case they share a qualitative part both among themselves and with that category. And two objects resemble with respect to a category just in case each shares a qualitative part with that category and these qualitative parts resemble each other.

The notions so far introduced would enable us to specify truth conditions of which the following is an informal example:

(4.4.7)   An atomic sentence of the form '*A* is *F*' is true just in case either
(1) '*F*' is a color predicate and *A* resembles, with respect to the category 'color', the designatum of '*F*', or
(2) '*F*' is a shape predicate and *A* resembles, with respect to the category 'shape', the designatum of '*F*', or ...

However, in the absence of a syntactical difference between color predicates, shape predicates, and so on, one would not know which clause of this definition is meant to apply to a given predicate '*F*'. Hence, the predicates of the object-language must be divided into as many syntactical classes as there are categories with respect to which distinct resemblance relations can obtain. Thus, we envisage an enumeration of all categories, of which c_i shall be the ith. Then we introduce into the vocabulary of the object-language, for every i, a non-schematic predicate 'R_i' which expresses resemblance in the ith respect, and a denumerable set of schematic predicates (perhaps bearing the distinctive subscript 'i') which may be unambiguously interpreted in the following style:

(4.4.8)   If 'F_i' is a predicate of the ith classification, then sentences of the form '*A* is F_i' are true just in case the designatum of '*A*' resembles, with respect to the ith category c_i, the designatum of 'F_i'.

The predicate 'R_i' in the object-language which expresses resemblance

BUNDLES OF QUALITIES, QUALITIES, AND CONCRETA 191

in the ith respect will serve in expressing a syntactical counterpart of (4.4.8).

Against those nominalists who seek to explicate classificatory statements in terms of the relation 'resembling with respect to c', the following objection is frequently urged: that statements of the form 'x resembles y with respect to c' imply the existence of a certain abstract entity, namely the respect c, so that nominalists who will not endorse abstract entities cannot consistently assert statements of that form, nor any classificatory statements whose analysis implies such statements. However, in interpreting the non-schematic predicates 'R_i' mentioned above (which express resemblance in the ith respect c_i), no assignment to 'R_i' need be employed due to which 'R_i' might designate the category c_i. For that reason, and recalling our criterion for ontological implications (Section 3.1), statements of the form '$x\, R_i\, y$' need not make actual reference to the category c_i, and hence will not imply the existence of a category. Generally, non-schematic predicates, of which 'R_i' is an example, need not enter referential assignments. If difficulties arise, they should be sought in the interpretation of *schematic* predicates.

There is, however, another difficulty. Supposing that we understand the expression 'x resembles y with respect to color' and that we have assigned to the color predicate 'is red' a concrete exemplar which is in fact red, it could be that the exemplar was so chosen that it is not only red but also, in other places, white and green. If so, one could not employ the exemplar in specifying the intended application of the predicate 'is red', since not only red things, but also green and white ones would resemble the exemplar with respect to color. Instead, the exemplar of 'is red' should be red all over; that is to say, it should be uniform with respect to color. Generally, the following definition serves to express what it means to be uniform in a given respect:

(4.4.9) DEFINITION: x is *uniform with respect to c* just in case c is a category, and for all y and z: if both y and z match x with respect to c, then y matches z with respect to c.

Thus, an object x is of uniform color, say red all over, if any two items which match x with respect to color will also match one another with respect to color. This definition, relativized to a given universe of discourse, serves to impose the condition that every exemplar which is assigned to a predicate of the ith classification shall be uniform with respect to the ith category. This condition (which we envisage as one imposed on assignments to schematic predicates in models) does not seem unduly restrictive; especially since we are free to regard any uniform portion of a concrete object itself as a concrete object. Hence, uniform parts of an object may serve as exemplars in the event that all conglomerates happen to lack uniformity.

Up to now we have roughly outlined how one-place predicates might be interpreted in a concretistic system. These are the predicates traditionally discussed in connection with nominalism. Of far greater interest, however, are predicates of higher degree. Indeed, if an adequate treatment of two-place predicates could be given, in the framework of a concretistic system, then the needed modifications for predicates of lesser and greater degree would probably become obvious. For this reason, we shall be concerned from now on only with relation expressions; indeed, we shall focus only on those which serve to express dyadic relations of qualitative comparison, like 'is warmer than', 'is darker than', 'is louder than', and the like.

In Section 3.4, a number of proposals were informally discussed which concerned the interpretation of relation expressions in a nominalistic system. One of them (the proposal (2) in that section) consisted in admitting the sequential individuals axiomatically treated in Section 2.10.4, and in assigning to an expression like 'warmer than' the sum of all sequential individuals (x, y) where x, intuitively, is warmer than y. Entering a relation could then be analysed as being a sequential individual in that sum. Thus, the relation of predication (in the slightly extended sense appropriate to relation expressions) would turn out to be definable in terms of the part-whole relation. The resulting system, by our present criteria, would be an abstractistic one. Already in Section 3.4, we have found independent grounds for judging that sequential individuals could not generally be regarded as concrete. The proposed interpretation of relation expressions, attractive as it would otherwise seem, can therefore not satisfy our present demand for a concretistic system.

Another proposal which has initial plausibility can be described as follows: To begin with, an auxiliary notion is required: that of 'degree of resemblance'. It can be recursively defined in the meta-theory as follows:

(4.4.10)	(1) x resembles y to the degree 0 just in case x R y;
(2) x resembles y to the degree $(n+1)$ just in case for some z, $z \neq y$, x R z, z resembles y to the degree n, and there exists no u such that x R u and for some $m < n$, u resembles y to the degree m.

Informally, we think of concentric resemblance-circles around a given quality y, where just qualities between two successive circles resemble one another. The degree of resemblance is given by the count of successively larger circles, starting at the center. In these terms, one can define a notion of comparative resemblance as follows:

(4.4.11)	*x resembles y more closely than z* just in case for some natural numbers $m < n$, x resembles y to the degree m and x resembles z to the degree n.

These definitions presuppose that there are no 'gaps' in any array of colors or other kinds of qualities. Already the previous definition of a 'clan' and of a 'category' are intuitive only under this assumption. However, the set of qualities, whose existence is postulated in the metatheory, is conceived as comprising all possible qualities (while in a given universe of discourse only occurrent qualities appear). Thus, every conceivable place in a resemblance chain may be regarded as occupied by one of these possible qualities.

Next, we require the notion: 'x resembles y more closely than z with respect to the category c'. It is clear how this notion is defined.

Given these concepts, one might proceed to interpret a relation expression, like 'is brighter than', much as we have previously described it in Section 3.5: One assigns to the ordered pair ⟨'is brighter than', the moon⟩ a concrete object which is just noticeably brighter than the moon. Then exactly those objects shall be said to be brighter than the moon which resemble that exemplar more closely than they resemble the moon. Thus, truth conditions for sentences of relational form could be precisely specified in the pattern of the following informal example:

(4.4.12)	The sentence '*A* is brighter than the moon' is true just in case the item named by '*A*' resembles, with respect to brightness, the exemplar assigned to ⟨'is brighter than', the moon⟩ more closely than the moon.

Although a formal semantical interpretation along these lines could easily be provided, a complete axiomatization of such a system presents the following serious difficulties: In all interpreted formal systems so far considered, one could express, in the object-language, a counterpart of every truth condition specified for atomic sentences in the meta-language. This was possible because there was a counterpart, in the object-language, of the relation of predication employed in the meta-language. In the system presently envisaged, the relation of predication is 'x resembles y with respect to c more closely than it resembles z'. Though various alternatives come to mind, it seems most natural to express a counterpart of that relation in the following way: As previously indicated, we enumerate all categories and divide the predicates into as many exclusive classes as there are categories. For every category c_i we introduce the primitive non-schematic predicate 'R_i' which is to express resemblance with respect to c_i. In order to formulate a counterpart of the relation 'x resembles y more closely than z with respect to c_i', it seems that one has to express the ancestral with respect to 'R_i'. That is to say, it appears that we need a formula $\phi(x, y)$ of the object-language of such a sort that

(4.4.13) $\phi(x, y)$ is true of the respective objects x and y exactly when x is in every class K which is such that y is in K and for every u and v, whenever v is in K and '$u\,R_i\,v$' is true of u and v then u is again in K.

However, in a first-order language, and in the indicated sense of 'expressing', it seems that one cannot express the ancestral with respect to an arbitrary relation R.

For suppose, to the contrary, that one could. Then it would be possible to determine that the universe of every model of a theory is exactly denumerable by specifying that R is a strict partial ordering, that for every x there is exactly one y such that xRy, and that there is an item x such that everything is an ancestor, with respect to R, of x. But this is impossible since, by the Upward Loewenheim-Skolem Theorem, every first-order theory which has an infinite model has a model which is more than denumerable.[7]

Since the approach just outlined does not look very promising, we shall pursue another one. We shall grant ourselves the assumption that certain ordered chains of concrete things are themselves concrete things; namely those in which concrete objects are lined up side by side. Within such a chain, we can distinguish individuals which are to the left of others. It is

BUNDLES OF QUALITIES, QUALITIES, AND CONCRETA   195

further supposed that all comparative qualitative relations, if they obtain at all, can be suitably exemplified by such chains of concreta. For example, we shall suppose that all discriminable degrees of brightness can be exemplified by a chain of concrete things which are lined up side by side in such a fashion that any two adjacent things in the chain resemble one another with respect to brightness, while each is uniform with respect to brightness. For the sake of simplicity, we shall also assume that there are no 'missing' degrees of brightness. For if we were to allow that some degrees are not occurrent in a given universe, we would have to face the additional difficulties (apparently overcome by Goodman, 1966) of adjusting linear arrays with missing resemblance links. Under these assumptions, we want to assign to the expression 'brighter than' an exemplifying chain of the sort described, where brighter things are to the left of less bright ones. It should then be true, of arbitrary objects x and y, that x is brighter than y just in case x resembles, with respect to brightness, an object x' in that chain, y resembles, with respect to brightness, some object y' in the chain, and x' is to the left of y'.

We shall presently outline, in somewhat greater detail, how we conceive of the semantics corresponding to that idea. Since it is merely intended as a tentative proposal, the development will fall short of complete rigor and no axiomatization is attempted. Throughout the discussion, adopt the convention that **U** is an atomistic universe of individuals for the part-whole relation **R**.

Let **L** be a diadic relation in **U**, where '**L**xy' is read 'x is to the left of y'. We define:

(4.4.14) y is an *immediate* **L**-*successor* of x just in case **L**xy and there is no z such that **L**xz and **L**zy; and x is an *immediate* **L**-*predecessor* of y if **L**xy and there is no z such that **L**xz and **L**zy.

The relation **L** shall satisfy the following conditions with respect to objects in **U**:

(4.4.15)   (1) If **L**xy then not: **L**yx,
(2) If **L**xy and **L**yz then **L**xz,
(3) If **L**xy, then there exist u and v such that u is an immediate **L**-successor of x and v is an immediate **L**-predecessor of y.

The conditions (1) and (2) provide that the relation **L** is a strict partial ordering, while (3) asserts that it is a discrete partial ordering.

As a further condition on **L** and **U** we would like to assert this: given any non-empty specifiable and discretely ordered set of atoms, the sum or chain composed of those atoms is again an individual. Although we shall not define the notion of a 'specifiable' set in the present context, we envisage an explication analogous to that given in connection with CIIII (Definition 2.7.1), but appropriate to the language yet to be described.

(4.4.16) Suppose that S is any non-empty specifiable set of individuals which are **R**-least in **U**. Assume further that for all x and y in S: (1) either **L**xy or **L**yx or $x = y$, and (2) if **L**xy then there exist in S an immediate **L**-successor of x and an immediate **L**-predecessor of y. Then there exists an individual z in **U** such that exactly those **R**-least elements of **U** bear **R** to z which are members of S.

Informally: Suppose that S is any non-empty discretely ordered set of atoms of such a sort that membership in S can be determined by some formula of the object-language. Then there exists a certain individual sum or chain whose ultimate constituents are just those in S.

Let us call those individuals 'chains' whose atoms are discretely ordered by **L**:

(4.4.17) z is a *chain* just in case for all **R**-least elements x and y in **U** which bear **R** to z the following conditions are satisfied: (1) either **L**xy or **L**yx or $x = y$, and (2) if **L**xy then there exist an immediate **L**-successor of x and an immediate **L**-predecessor of y both of which are **R**-least elements of **U** which bear **R** to z.

We shall assume that our theory is adequate for expressing the relations of matching and resembling with respect to a finite number **n** of distinct categories. Accordingly, we divide the schematic two-place predicates of the object-language into **n** infinite and disjoint categories. An assignment **P** to predicates will have the following characteristics:

(4.4.18) **P** is a function whose domain is a set of two-place schematic predicates, and if π_i is any predicate in its domain which

BUNDLES OF QUALITIES, QUALITIES, AND CONCRETA 197

belongs to the ith category, then $\mathbf{P}(\pi_i)$ is a member of $\mathbf{U}$ which is a chain and such that (1) for every $\mathbf{R}$-least element x of $\mathbf{U}$ which bears $\mathbf{R}$ to $\mathbf{P}(\pi_i)$ and for every y and z in $\mathbf{U}$, if both y and z match x with respect to the ith category, then y matches z with respect to the ith category, and (2) for all $\mathbf{R}$-least elements x and y of $\mathbf{U}$ which bear $\mathbf{R}$ to $P(\pi_i)$, if x is an immediate $\mathbf{L}$-predecessor of y then x resembles y with respect to the ith category.

Thus, each two-place predicate of the ith category is to designate a certain concrete chain of concrete atoms of such a sort that (1) each atom in the chain is uniform in the ith respect and (2) adjacent atoms in the chain resemble one another with respect to the ith category. In these terms one can specify truth conditions for sentences of relational form in roughly the following pattern:

(4.4.19) If π_i is a two-place predicate belonging to the ith category, then π_i applies to the respective individuals x and y in $\mathbf{U}$ just in case there exist $\mathbf{R}$-least elements x' and y' of $\mathbf{U}$ which bear $\mathbf{R}$ to $P(\pi_i)$ such that (1) x matches x' with respect to the ith category, (2) y matches y' with respect to the ith category, and (3) $\mathbf{L}x'y'$.

Thus, the expression 'is brighter than' might be interpreted as follows: we assign to that expression a certain concrete chain of individuals each ultimate element of which is uniform with respect to brightness and such that any two adjacent elements in the chain resemble one another with respect to brightness. Let the brighter things in the chain stand to the left of less bright ones. Then of arbitrary concrete individuals x and y it can be said that x is brighter than y if x matches (with respect to brightness) some concrete atom x' in the chain, y matches (in the same respect) some atom y' in the chain, and x' is to the left of y'.

In this way, it seems, one can interpret arbitrary relation expressions, provided one presupposes, in addition to concepts of similarity, at least one ordering relation (such as 'to the left of') which is expressed by a non-schematic predicate.

4.5. FURTHER PROBLEMS AND CONCLUSION

We have set ourselves the task of explicating the notion of an individual and of interpreting formal theories according to the demands of various nominalistic positions. Due to our limitations of energy and ingenuity, numerous problems were left out of the discussion which have traditionally been of concern to nominalists. Among them, we mention especially the following:

Throughout the present work, we have employed a set theoretic meta-theory. Thus, in effect, the syntax and semantics of nominalistic systems has been described in the framework of a theory which is not itself nominalistic. Additional problems arise if nominalistic systems are to be characterized in a meta-theory which is itself acceptable to nominalists. For example, while one may safely assume, in a set theoretic meta-language, that there exist infinitely many items which can serve as symbols of the object-language, no such assumption seems warranted if all expressions of the language are to be concrete individuals. We have not even begun to solve such problems which beset 'meta-nominalism'.

One of the most urgent problems, for a nominalist, is that of expressing those numerical relations which seem to be required in the sophisticated sciences. The problem is at least three-fold: (1) there is the problem of determining in general exactly which numerical relations a nominalist would want to express. Thus, a nominalist might not want to endorse a system from which it follows that there is even one number, assuming that numbers cannot somehow be identified with individuals. For this reason, he may not even wish to express (or, if he does, he is likely to deny) statements like 'there exists a number between 2 and 4'. On the other hand, a nominalist would presumably want to express certain assertions of cardinal comparison, such as 'there are fewer dogs than cats', provided they are so interpreted that their truth does not imply the existence of such items as cardinal numbers or functions. But it has never been determined, as far as I know, exactly what sort of numerical assertions a nominalist would want to express. (2) There is the problem of deciding just what exactly shall be meant by 'expressing' a numerical relation; and (3), assuming a clarification of the first two points, there is the problem of how one 'expresses' (in the given sense) exactly those quantitative assertions which a nominalist finds comprehensible.[8]

Another large area, of concern to nominalists, is that of providing an acceptable theory of meaning which avoids reference to presumed non-individuals such as concepts, senses, properties, or 'meanings'. While some interesting suggestions have been made [9], no precise and comprehensive theory of meaning is known which accords with nominalism.

While nominalism presents numerous important problems whose precise formulations and solutions had to be left out of the present account, we hope to have illustrated how the method of formal theory construction can profitably be employed in investigating at least some of the problems which have traditionally been of concern to nominalists.

REFERENCES

[1] Russell (1946), (1940), (1948).
[2] For details of a similar proof, see, e.g., Eberle (1969b).
[3] A treatment of properties within intensional logic is given, e.g., in Kaplan (1964).
[4] Goodman (1966), especially pp. 145–9.
[5] Implausibly strong conditions would have to be imposed on universes of discourse in order to avoid the difficulties of 'companionship' and of 'imperfect community' mentioned by Goodman (1966).
[6] Goodman (1966), pp. 286–7.
[7] For a general formulation of the Upward Loewenheim-Skolem Theorem, see Tarski and Vaught (1957). The use of this theorem in the given context was suggested to me in discussion by David Kaplan.
[8] See especially, Goodman and Quine (1940), Martin (1943), (1949), and Henkin (1953), (1960).
[9] For example, by Goodman (1952).

BIBLIOGRAPHY

Aaron, J. R.: 1952, *The Theory of Universals*, Oxford.
Anderson, A. R.: 1957, Review of Cartwright (1954), in *The Journal of Symbolic Logic* **22**, 393–4.
Benacerraf, P. and Putnam, H. (eds.): 1964, *Philosophy of Mathematics, Selected Readings*, Englewood Cliffs, N.J.
Beth, E. W.: 1959, *The Foundations of Mathematics*, Amsterdam.
Bocheński, J. M.: 1956, *The Problem of Universals. A Symposium*, Notre Dame, Ind., pp. 33–57. (With contributions by A. Church and N. Goodman.)
Carnap, R.: 1947, *Meaning and Necessity*, Chicago. Second ed., with supplements, 1956.
–: 1950, 'Empiricism, Semantics, and Ontology', *Revue International de Philosophie* **11**. Reprinted in numerous collections; in particular also in (1947), 2nd ed.
–: 1966, *Philosophical Foundations of Physics* (ed. by M. Gardner), New York-London.
Cartwright, R. L.: 1954, 'Ontology and the Theory of Meaning', *Philosophy of Science* **21**, 316–25.
Church, A.: 1951, 'The Need for Abstract Entities in Semantic Analysis', *Proceedings of the American Academy of Arts and Sciences* **80**, 100–2.
–: 1958, 'Ontological Commitment', *The Journal of Philosophy* **55**, 23, 1008–15.
Chwistek, L.: 1924, 'The Theory of Constructive Types', *Rocznik Polskiego Towarzystwa Matematycznego* [Annales de la Société Polonaise de Mathématique] **2**, 9–48; **3**, 92–141.
Clay, R. E.: 1965, 'The Relation of Weakly Discrete to Set and Equinumerosity in Mereology', *Notre Dame Journal of Formal Logic* **6**, 4, 325–40.
–: 1968, 'The Consistency of Lesniewski's Mereology Relative to the Real Number System', *The Journal of Symbolic Logic* **33**, 2, 251–7.
Copi, I. M.: 1958, 'Objects, Properties and Relations in the "Tractatus"', *Mind* **67**, 145–65.
Copi, I. M. and Gould J. A. (eds.), 1967, *Contemporary Readings in Logical Theory*, New York and London.
Eberle, R. A.: 1965, *Nominalistic Systems – The Logic and Semantics of some Nominalistic Positions*. Dissertation, The University of California at Los Angeles.
–: 1967, 'Some Complete Calculi of Individuals', *Notre Dame Journal of Formal Logic* **8**, 4, 267–78.
–: 1968, 'Yoes on Non-Atomic Systems of Individuals', *Nous* **2**, 4, 399–403.
–: 1969a, 'Universals as Designata of Predicates', *American Philosophical Quarterly* **6**, 2, 151–7.
–: 1969b, 'Denotationless Terms and Predicates Expressive of Positive Qualities', *Theoria* **35**, part 2, 104–23.

–: 1969c, 'Non-Atomic Systems of Individuals Revisited', *Nous* **3**, 4, 431–434.
Gödel, K.: 1931, 'Über formal unentscheidbare Sätze der *Principia Mathematica* und verwandter Systeme', *Monatshefte für Mathematik und Physik* **38**, 173–98.
Goodman, N.: 1949, 'On Likeness of Meaning', *Analysis* **10**, 1–7. Reprinted in Linsky (1952).
–: 1951, *The Structure of Appearance*, Cambridge, Mass.
–: 1955, *Fact, Fiction and Forecast*, Cambridge, Mass. (2nd ed.: Indianapolis 1965).
–: 1956, 'A World of Individuals', in *The Problem of Universals. A Symposium*, Notre Dame, Ind., pp. 13–31. (With contributions by A. Church and J. M. Bocheński). Reprinted in Copi and Gould (1967).
–: 1958, 'On Relations that Generate', *Philosophical Studies* **9**, 65–6.
–: 1966, *The Structure of Appearance*, 2nd revised ed., Indianapolis–New York–Kansas City.
Goodman, N. and Quine, W. V. O.: 1947, 'Steps toward a Constructive Nominalism', *The Journal of Symbolic Logic* **12**, 105–22.
Halmos, P. R.: 1963, *Lectures on Boolean Algebras*, Princeton, N.J., New York.
Hellman, G.: 1969, 'Finitude, Infinitude, and Isomorphism of Interpretations in some Nominalistic Calculi', *Nous* **3**, 4, 413–425.
Hempel, C. G.: 1953, 'Reflections on Nelson Goodman's "The Structure of Appearance"', *Philosophical Review* **62**, 108–16.
–: 1957, Review of Goodman (1956), in *The Journal of Symbolic Logic* **22**, 205–8.
Henkin, L.: 1953, 'Some Notes on Nominalism', *The Journal of Symbolic Logic* **18**, 19–29.
–: 1962, 'Nominalistic Analysis of Mathematical Language', in *Logic, Methodology, and Philosophy of Science. Proceedings of the* 1960 *International Congress* (ed. by Nagel, Suppes, and Tarski), pp. 187–93.
Hintikka, J.: 1959, 'Existential Presuppositions and Existential Commitments', *The Journal of Philosophy* **56**, 125–37.
Hodges, W. and Lewis, D.: 1968, 'Finitude and Infinitude in the Atomic Calculus of Individuals', *Nous* **2**, 4, 405–10.
Kalish, D. and Montague, R.: 1964, *Logic – Techniques of Formal Reasoning*, New York and Burlingame.
Kaplan, D.: 1964, *Foundations of Intensional Logic*. Dissertation, University of California at Los Angeles.
Kneale, W. and M.: 1962, *The Development of Logic*, Oxford.
Küng, G.: 1967, *Ontology and the Logistic Analysis of Language*, Dordrecht, Holland.
Leonard, H. S.: 1930, *Singular Terms*. Dissertation, typescript, Widener Library.
Leonard, H. S. and Goodman, N.: 1940, 'The Calculus of Individuals and its Uses', *The Journal of Symbolic Logic* **5**, 45–55.
Leśniewski, St.: 1927–31, 'O Podstawach Matematyki', *Przeglad Filosoficzny* **30** (1927), 164–206; **31** (1928), 261–91; **32** (1929), 60–101; **33** (1930), 77–105; **34** (1931), 142–70.
Linsky, L. (ed.): 1952, *Semantics and the Philosophy of Language*, Urbana, Ill.
Lowe, V.: 1953, 'Professor Goodman's Concept of an Individual', *Philosophical Review* **62**, 117–26.
Luschei, E. C.: 1962, *The Logical Systems of Leśniewski*, Amsterdam.
Martin, R. M.: 1943, 'A Homogeneous System for Formal Logic', *The Journal of Symbolic Logic* **8**, 1–23.
–: 1949a, 'A Note on Nominalism and Recursive Functions', *The Journal of Symbolic Logic* **14**, 27–31.

—: 1949b, 'A Note on Nominalistic Syntax', *The Journal of Symbolic Logic* **14**, 226–7.
—: 1958, *Truth and Denotation*, Chicago.
Mates, B.: 1965, *Elementary Logic*, Oxford.
Mendelson, E.: 1964, *Introduction to Mathematical Logic*, Princeton–Toronto–New York–London.
Moody, E. A.: 1953, *Truth and Consequence in Medieval Logic*, Amsterdam.
Myhill, J. R.: 1950, Review of Martin (1949a) and Martin (1949b). In *The Journal of Symbolic Logic* **15**, 153.
Quine, W. V. O.: 1940, *Mathematical Logic*, Harvard.
—: 1943, 'Notes on Existence and Necessity', *The Journal of Philosophy* **40**, 113–27. Reprinted in Linsky (1952).
—: 1947, 'On Universals', *The Journal of Symbolic Logic* **12**, 74–84.
—: 1949, 'Designation and Existence', *The Journal of Philosophy* **36**, 701–9.
—: 1951, 'On Carnap's Views on Ontology', *Philosophical Studies* **2**, 65–72.
—: 1951a, 'Two Dogmas of Empiricism', *The Philosophical Review* **60**, 20–43. Reprinted in Quine (1953).
—: 1951b, 'On What There Is', *Symposium III, Aristotelean Society*, Supplementary vol. XXV, pp. 149–60.
—: 1951c, 'Semantics and Abstract Objects', *Proceedings of the American Academy of Arts and Sciences*.
—: 1953, *From a Logical Point of View*, Cambridge, Mass.
—: 1960, *Word and Object*, Cambridge, Mass.–New York.
—: 1966, *The Ways of Paradox and other Essays*, New York.
Ramsey, F. P.: 1925, 'Universals', *Mind* **34**, 401–17. Reprinted in *The Foundations of Mathematics*, London 1931.
Routley, R.: 1966, 'Some Things do not Exist', *Notre Dame Journal of Formal Logic* **7**, 3, 251–76.
Russell, B.: 1912, 'On the Relation of Universals and Particulars', *Proceedings of the Aristotelean Society* **12**, 1–24. Reprinted in Russell (1956).
—: 1940, *An Inquiry into Meaning and Truth*, New York–London.
—: 1946, 'Problem of Universals', *Polemic* **2**.
—: 1948, *Human Knowledge, its Scope and Limits*, New York.
—: 1956, *Logic and Knowledge, Essays 1901–1950* (ed. by R. C. Marsh), New York.
Schuldenfrei, R.: 1969, 'Eberle on Nominalism in Non-Atomic Systems', *Nous* **3**, 4, 427–430.
Sellars, W.: 1960, 'Grammar and Existence: a Preface to Ontology', *Mind* **69**.
Sikorski, R.: 1964, *Boolean Algebras*, Berlin–Göttingen–Heidelberg–New York.
Sinisi, V. F.: 1962, 'Nominalism and Common Names', *Philosophical Review* **71**, 233f.
Sobociński, B.: 1955, 'Studies in Leśniewski's Mereology', *Rocznik Polskiego Towarzystwa Naukowego na Obczyźnie* **5**, 34–43.
Stegmüller, W.: 1956–57, 'Das Universalienproblem einst und jetzt', *Archiv für Philosophie* **6** (1956) 192–225; **7** (1957) 45–81.
Strawson, P. F.: 1959, *Individuals – An Essay in Descriptive Metaphysics*, London.
Suppes, P.: 1957, *Introduction to Logic*, Princeton, N.J.
—: 1960, *Axiomatic Set Theory*, Princeton, N.J.
Tarski, A.: 1936, 'Der Wahrheitsbegriff in den formalisierten Sprachen', *Studia Philosophica* **1**, 261–405. Reprinted in Tarski (1956).
—: 1944, 'The Semantic Conception of Truth', *Philosophy and Phenomenological Research* **4**, 341–75. Reprinted in Linsky (1952).

–: 1956, *Logic, Semantics, Metamathematics*, Oxford.
Tarski, A. and Vaught, R. L.: 1957, 'Arithmetical Extensions of Relational Systems', *Compositio Mathematica* **13**, 81–102.
Turbayne, C. M.: 1968, 'The Subject-Predicate Myth', *Studies in the 20th Century* **1**, 1, 7–20.
Woodger, J. H.: 1937, *Axiomatic Method in Biology*, Cambridge–New York.
Yoes, M. G., Jr.: 1967, 'Nominalism and Non-Atomic Systems', *Nous* **1**, 2, 193–200.

SYMBOLIC NOTATION

Notation	Reading	Page
$x = y$	x is the *same* as y	11, 45
$x \in y$	x is a *member* of y	12
$\{x : \phi\}$	the set of all x such that ϕ	12
0	the *empty set*, zero	12
$x \subseteq y$	x is *included* in y	12
$x \subset y$	x is *properly included* in y	12
$x \cup y$	the *union* of x and y	12
$x \cap y$	the *intersection* of x and y	12
$x \sim y$	the *difference* between x and y	12
$\{x\}$	singleton x	12
$\{x, y\}$	doubleton x, y	12
$\{x_0, \ldots, x_{n-1}\}$	the set whose members are $x_0, \ldots, x_{n-1}$	13
$\bigcup x$	the *union* of x	13
$\bigcap x$	the *intersection* of x	13
(x, y)	the *ordered pair* x, y	13, 82
$(x_0, \ldots, x_{n-1})$	the n-tuple whose respective constituents are $x_0, \ldots, x_{n-1}$	83, 97
xRy	x bears R to y	13
$\check{R}$	the *converse* of R	13
R_S	R *restricted* to S	14
Id_S	*identity restricted* to S	14
$\subseteq_S$	*inclusion restricted* to S	14
$f(x)$	the *value* of f at x	14
f is 1-1	f is a *one-to-one* function	15
f/g	the *relative product* of f and g	15
$\langle x \rangle$	the 1-term sequence x	15
$\langle x, y \rangle$	the 2-term sequence x, y	15
$\langle x_0, \ldots, x_{n-1} \rangle$	the n-term sequence whose respective constituents are $x_0, \ldots, x_{n-1}$	15
x_i	the value of x at i	15

Notation	Reading	Page
$\sim$	the negation sign	17
$\rightarrow$	the conditional sign	17
&	the conjunction sign	17
$\vee$	the disjunction sign	17
$\leftrightarrow$	the biconditional sign	17
$\forall$	the universal quantifier	17
$\exists$	the existential quantifier	17
$\ulcorner \xi_0, ..., \xi_{n-1} \urcorner$	the *concatenate* of the respective expressions $\xi_0, ..., \xi_{n-1}$	18
α, β, γ	variables ranging over variables	18
ϕ, ψ, χ	variables ranging over formulas	18
τ, ζ, η	variables ranging over terms	18
o, δ	variables ranging over operation symbols	18
π	variable ranging over predicates	18
ϕ^α_τ	the *proper substitution* of τ for α in ϕ	18
$\phi^{\alpha_0 ... \alpha_{n-1}}_{\tau_0 ... \tau_{n-1}}$	the *proper substitution* of $\tau_0, ..., \tau_{n-1}$ respectively for $\alpha_0, ..., \alpha_{n-1}$ in ϕ	18
K	variable ranging over sets of formulas	19
$\langle V, L \rangle$	the formal system whose vocabulary is V and whose proper axioms are in L	19
$\mathbf{U}$	a *universe of discourse*	20, 42, 76, 144
$\mathbf{O}$	an *assignment to operation* symbols	20, 144
$\mathbf{P}$	an *assignment to predicates*	20, 144
$\mathbf{A}$	an *assignment to variables*	20, 56
$\mathbf{M}$	a model	20, 56, 144
$\langle \mathbf{U}, \mathbf{O}, \mathbf{P}, \mathbf{A} \rangle$	a *classical model*	20
$\sup_\mathbf{R} S$	the *supremum*, relative to $\mathbf{R}$, of S	32
$\mathbf{R}$	a *part-whole relation*	33
$x \circ y$	x overlaps y	45
$x \leq y$	x is part of y	45
$x < y$	x is a proper part of y	45
$At x$	x is an atom	45
$\langle \mathbf{R}, \mathbf{U}, \mathbf{A} \rangle$	a model of CI I	56
$\mathbf{A}^\alpha_x$	that assignment which differs from $\mathbf{A}$, if at all, only by assigning x to α	56
$\mathbf{M}$ sat ϕ	$\mathbf{M}$ satisfies ϕ	57, 152

Notation	Reading	Page
$inf_R(x, y)$	the *infimum*, relative to R, of x and y	63
$Cxyz$	x is the concatenate of y and z	85
$x+y$	the sum of x and y	85
$x\,\mathbf{L}\,y$	x is immediately to the left of y	87
Wxy	x is (together) with y	90
$\mathbf{l}$	the left position	92
$\mathbf{r}$	the right position	92
$Lxyz$	y links x to z	95
$\langle \mathbf{R}, \mathbf{U}, \mathbf{O}, \mathbf{P}, \mathbf{A} \rangle$	a *nominalistic model*	144
$x\,M\,y$	x matches y	150
$Den(\mathbf{M}, \tau, x)$	in $\mathbf{M}$, τ denotes x	152
$x\,R\,y$	x resembles y	168
$\mathbf{N}$	a neighborhood assignment	171
Σx	x is a set	178
$x \leq^* y$	x is part of y	179
$x <^* y$	x is a proper part of y	179
$x\,R_i\,y$	x resembles y in the ith respect	190
$\mathbf{L}xy$	x is to the left of y	195

INDEX OF NAMES

Aaron, J. R. 200
Abelard, Peter 4
Anderson, A. R. 200
Anselm of Canterbury 4
Aristotle 4

Benacerraf, P. 200
Berkeley, George 6, 148
Beth, E. W. 200
Bocheński, J. M. 147, 200

Carnap, R. 140, 200
Cartwright, R. L. 111–112, 114, 147, 200
Champeaux, William of 4
Church, A. 105, 147, 200
Chwistek, L. 7, 200
Clay, R. E. 99, 200
Copi, I. M. 200

Eberle, R. A. 100, 101, 147, 199, 200, 201

Frege, G. 143

Gödel, K. 201
Goodman, N. 7, 24–31, 32–33, 35–36, 39–42, 44, 48, 49, 73, 77, 82, 90, 91, 99–101, 170, 178, 185–186, 195, 199, 201
Gould, J. A. 200

Halmos, P. R. 99, 201
Hellman, G. 201
Hempel, C. G. 81, 99, 101, 201
Henkin, L. 7, 199, 201
Hintikka, J. 201
Hobbes, Thomas 6
Hodges, W. 201
Horn, A. v, 100
Hume, David 6

Kalish, D. v, 23, 147, 201

Kaplan, D. v, 100, 147, 199, 201
Kneale, W. and M. 201
Küng, G. 201

Leonard, H. S. 44, 100, 201
Leśniewski, S. 7, 99, 201
Lewis, D. 201
Linsky, L. 201
Locke, John 6
Lowe, V. 201
Luschei, E. C. 100, 201

Martin, R. M. 7, 124, 147, 199, 201, 202
Mates, B. 23, 202
Mendelson, E. 100, 202
Montague, R. v, 23, 147, 201
Moody, E. A. 202
Myhill, J. R. 7, 202

Ockham, William of 4

Plato 3–4
Porphyry 4
Putnam, H. 200

Quine, W. V. O. 7, 23, 82, 99, 102–109, 111, 114, 124, 126, 143, 147, 199, 202

Ramsey, F. P. 202
Roscelin 4
Routley, R. 144, 147, 202
Russell, B. 100, 122, 143, 147–148, 199, 202

Schuldenfrei, R. 100, 202
Scotus, John 4
Sellars, W. 202
Sikorski, R. 100, 202
Sinisi, V. F. 202
Sobociński, B. 99, 202
Stegmüller, W. 202

Strawson, P. F. 202
Suppes, P. 16, 178, 202

Tarski, A. 7, 99, 125, 147, 199, 202, 203
Turbayne, C. M. 203

Vaught, R. L. 147, 199, 203

Woodger, J. H. 7, 99, 203

Yoes, M. G., Jr. 73, 74, 100, 203
Yost, R. M. v

INDEX OF SUBJECTS

Abstract entities 8, 9, 99, 105
—, individuals 24, 91, 97, 99, 133, 147, 148, 175, 184–188, 192
—, *see* also abstraction, abstractistic
Abstraction 4, 9
—, problem of 186–187
Abstractistic system 186, *187*, 188, 192
Actualized in a universe 38–39, 90–92, 123, 128, 139, 193
Affirmed in a theory 105–106
Algebra, *see* Boolean algebra
Analysis (philosophical) 1
Ancestor with respect to a relation *16*
Ancestral 16
—, of membership *16*, 27–28, 73–74, 100
—, how to express 194
Antisymmetric relation *14*
Arithmetic, *see* mathematics
Assignment of neigborhoods 171
—, referential 115, 116
—, to operation symbols 20, *145*, *146*
—, to predicates 20, *145*–146, 196–197
—, to variables 20, *56*, 108, 115, *145–146*, 152, 171
Asymmetric relation *14*
Atom 27, 30–31, *33*, *46*, 48–50, 73–81, 83–98, 176, 179–181
—, *see* also R-least, atomic, atomistic
Atomic individual, *see* atom
—, formula 45, 95
—, link 93–98
Atomistic calculus of individuals 30, 49–73, 94–99
—, universe 36–44, *42*, 49–50, 93–94, 144–145, 151
—, *see* also non-atomistic
Axiom, definitional 20
—, logical 19, *45*
—, of the meta-theory 11, 178–181, 189
—, of LGCI 45–47

—, of CI I 50–51
—, of CI II 63
—, of CI III 69
—, of CI IV 78–79
—, of the calculus of individual relations 95–96, 98
—, of CB 158–160
—, of CR 168–170
—, of CQ 181
—, of Regularity 11, 100
—, proper, of a formal system 19

Boolean algebra, and part-whole relations 36, 99
—, and the calculus of individuals 49, 78, 100
—, ideals 100
Bound terms 18
Bundles as sets of individuals 151
—, being 'in' a bundle 149, 151
—, calculus of (CB) 150–166
—, designated by predicates 132–133, 154–156, 186
—, individuated 149, 151, 168–169
—, of qualities 3, 7, 148–166, 168–170, 171, 176–177
—, pre-systematically 148–149, 151, 154–155, 168–169

Category *189*, 193
Calculus of individuals 1, 8, 24, 28, 44–81, 99–100
—, CI I 50–62, 100
—, CI II 62–67, 100
—, CI III 67–73, 100
—, CI IV 77–81
—, CB (bundles of qualities) 150–166
—, CR (resemblance) 168–175
—, CQ (qualities) 181–184
—, individual relations 92–99
—, Leonard and Goodman 44–50, 69,

77–78, 100
—, *see also* mereology
Cardinality of a set 15
Chain 194–195, *196*, 197
Clan *189*, 193
Class, *see* set
—, vs. individual, *see* individual, nominalism
Classical model *20*, 21–23, 114–117, 123, 136, 142
Classification 3, 5, 134
Closure under a relation *16*
—, under an operation *16*
—, under modus ponens *19*
—, under universal generalization *19*
—, under sums 40–41, 50, 62–73, 91, 98
—, under products 50, 62–73
—, under difference 78
Collection, *see* set, sum
Companionship difficulty 199
Comparative concepts 140, 192
Comparison 135, 137, 138–139, 166–167, 169, 192
Completeness *23*, 44, 100, 114, 126
—, of CI I 62
—, of CI II 66–67
—, of CI III 71–73
—, of CI IV 80–81
—, of CB (bundles of qualities) 163–166
—, of CR (resemblance) 173–175
—, of CQ (qualities) 184
—, of the calculus of individual relations 96
Concatenation 18, 82–83, 85–89, 97
Concept 5, 8
Conceptualism 4, 10
Concrete individuals 7, 8, 24–25, 28–31, 90, 92, 96, 98, 103, 124, 133–134, 137–139, 147, 148, 175, 184–197
Concretistic system 186, *187*, 188, 192
Concretum 90, 184, 185–186
Connective *17*
Consistent *19*
—, maximally *19*
Constant 17, 22
Constituent, *see* part
Constructionism 1
Construing 109–110

Content 26, 30–31, 82
Continuum 15, 178
Converse relation *13*
Corners *18*
Criterion, *see* ontological

Deduction theorem 58, 161
Defined symbols *20*, 97
—, in LGCI 45
—, in CI I 50
—, in CB 150
Definition *20*
Degree of resemblance 192, *193*
Denial by a theory 106, 119
Denotation in models 22, 142–147, *152*, 182
—, *see also* designation
Derivation *19*
Description, definite *11–12*, 20, 100, 106, 119, 143
Designation, and ontological implications 112–117
—, in models 22, 142–147, 152–155, 182
—, multiple 5, 124, 127–130, 136–137
—, non-designating predicates 4–5, 20–21, 113, 123–142, 144–147, 156
—, non-designating terms 2, 112, 142–147, 153–155
—, of predicates 4–5, 20–21, 113, 123–142, 144–147, 153–156, 186–188, 190, 192, 195–197
Difference *12*
—, closure under 78
—, noticeable 149
Discrete individuals 34–35, 50, 64, 77, 90, 189
—, ordering 196
Display 18
Distinguishing 166–167, 182
Domain of a relation *13*
Doubleton *12*

Empty set *7*
—, and nominalism 7, 35, 180
—, designated by improper descriptions 12, 32, 34
Equinumerous *15*
Equivalence relation *14*

INDEX OF SUBJECTS 211

—, class *14*
Essence 4
Exemplar 124, 131, 133–137, 139, 142, 146, 186–188, 191, 195, 197
Existence, and the empty universe 46, 56, 100, 143, 145
—, and denotationless terms 86, 112, 143, 153
—, expressed by a predicate 46–47, 86, 113, 158–159
—, implied by positive qualities 155–156
—, implied in a theory, *see* ontological
—, of an infimum *63*
—, of a set *12*
—, of a supremum *32*
—, *see* also closure, sum, individual
Extension of a predicate 21, 114, 115, 117, 119, 122–123, 125, 131
—, assignment *20*, 123, 136, 145–146, 154–156
Extensional context 11, 153, 175
Extensionality 26, 151

Fewer *15*, 198
Field of a relation *13*
Finite *15*
—, sums and products 62–67
Finitism 8, 30, 39, 141
Form (Platonic) 3–4
Formal system, *see* system
Formula *18*
—, of a formal system *19*
Fragmentation 29–30
Free terms *18*
Function *14-15*
—, value *14*
—, one-to-one *15*
Fusion, *see* sum

General ideas 6
—, names 4, 6
—, objects 7
Generating relation 26–29, 31, 73–74
—, *see* also part-whole relation
Greek letters *18*, 178

Heap 25, 28–29

Idea, for Berkeley, *see* quality

—, for Plato 3–4
—, general and particular 6
Identity *11*, *14*
—, as a relation of predication 127–128
—, in CB 158
—, in CI I 50
—, in CI IV 78–79
—, in CR 168
—, in LGCI 45, 100
—, of sequential individuals 97
—, *see* also individuation
Immediate L-successor *195*
Immediately to the left of 87–89
Imperfect community 199
Implication, ontological 102–122
Independence result 48–49
Inclusion *12*, *14*
Individual, abstract 24, 91, 97, 99, 133, 147, 148, 175, 184–188, 192
—, actual vs. possible 38, 182, 193
—, and class 24, 27, 44, 99, 128, 147, 148, 175
—, and heaps 25, 28–29
—, as item in space and time 9, 151
—, as member of the universe of discourse 38–39, 56, 92, 123–124, 128, 146, 178
—, as nameable item 4, 68, 123, 124, 128
—, as non-set 147, 148, 175, 177–178
—, as set 99, 147, 148, 151
—, atomic, *see* atom, atomistic, R-least
—, calculus of, *see* calculus
—, chain 194–195, *196*, 197
—, characterized in terms of totalities 36–37
—, concatenates 82–83, 85–89, 97–99
—, concrete, *see* concrete
—, determined by content 25–31, 81–82
—, for Aquinas 5, 31
—, for Aristotle 4, 5
—, for Goodman 24–33, 36, 39, 40–41, 81–82, 84, 90–92, 99, 185–186
—, perceived or given 9, 25, 149
—, pre-systematically 30–31, 81–82, 185
—, relational 92–99, 138–139

—, repeatable 8, 97
—, scattered 24–25, 40–41, 138
—, sequential or ordered 28–30, 81–99, 137–139, 192
—, string 97, 98
—, sum, *see* sum
—, universal 70
—, universe of, *see* universe
Individuation, principles of 2, 5, 25–31, 36, *37*, 38–39, 48–49, 55, 73, 76–77, 82–83, 89–90, 97, 149, 151, 168–170, 178
Inference rule *19, 47*, 161
Infimum *63*
Infinite collection *15*
—, and nominalism 8, 30, 39, 141, 198
—, in the meta-theory 11, 35, 179, 198
Inscription 8, 85, 198
Interpretation, model-theoretic 23
—, *see* also model, predicate, operation symbol
Intersection *12, 13*
Irreflexive relation *13*
IR-relation *93*
Isomorphic models *182–183*, 184
—, relations *16*, 177

Justification of classifications 5
—, of truth claims 128–130

Kind 104–105, 110–112, 115–116, 128
Knowledge, objects of 5

Leibniz' Law 45, 59, 159
Left 90–92
—, immediately to the left of 87–89, 99, 195–196
—, to the left of 195–197
Link 93, *94*, 95–99, 138
Linking relation 95–99
Loewenheim-Skolem Theorem 194, 199
Logical axiom *19, 45*
—, constant *22*
—, inference rule *19*

Manor 170
Matching 1, 150, 158–160, 170, 176–181
—, as a relation of predication 157

—, as overlapping 150, 157
—, as partial qualitative identity 157, 159, 166, 176
—, in a given respect *190*
—, in the meta-theory 178–181
—, interpreted in CB 152, 157
—, vs. resembling 157, 167, 170
   see also resemblance
—, the symbol of 150, 178
Material object 7, 185
Mathematics and nominalism 8, 82, 198
Maximally consistent *19*
Meaning 8, 122, 129–130, 198
—, theory of 8, 198
Membership 12, 25
—, as a relation of predication 127–128, 130
—, chains *16*, 26–27, 73–74, 100
Mereology 7, 99
Meta-theory *11*, 17–23, 126, *177–181*, 189, 198
—, not nominalistic 11, 44, 178–179, 198
—, provable in 110–111, 116–122
Mind-dependence of universals 4
Minimal element *14*
—, *see* also R-least element
Model 20–22
—, and realism 123, 136, 142
—, as 'possible world' 38–39, 99
—, classical *20*, 21, 114, 115, 117, 123, 136, 142
—, denotation in *22*, 142, 147, *152*
—, for individual concatenates 87
—, isomorphic *182–183*
—, nominalistic 142–143, *144–145*, 146–148, 151, 171, 181
—, of CB (bundles of qualities) *151*
—, of CI I *56*
—, of CI II *64*
—, of CI III *70*
—, of CI IV *80*, 81
—, of CQ (qualities) *181*
—, of CR (resemblance) *171*
—, of the calculus of individual relations *95*
—, satisfaction in *22*, 57, 89, *95*, *152–153, 172*
—, standard 102–103

—, truth in *22*, 147, *153*
—, value in *22*, 144
Multiple designation 5, 124, 127–130, 136–137

Name 17, 21, 153
—, and ontic import 106, 112
—, universal 4, 6, 123
—, vs. predicate 123–124, 128
—, *see* also term, description, designation
Neighborhood assignment *171*
Nominalism, advantages of 128, 130–131, 133, 135
—, and abstract items 9, 99, 133, 138–139, 184, 186–187, 191
—, and British empiricism 6, 148–150
—, and fragmentation 29–30
—, and parsimony 6, 8, 128
—, as an ontological doctrine 3, 124–125
—, avoiding classes and relations 6, 8–9, 44, 99, 128, 147
—, avoiding 'general objects' 7
—, avoiding intensions 8, 10
—, avoiding non-individual designata of predicates 2, 123–124, 128, 136, 146
—, avoiding non-individual designata of relation- and operation expressions 135–142
—, avoiding possibilia 39
—, avoiding postulated infinities 8, 30, *39*, 141, 178, 198
—, avoiding reference to non-individuals 1, 102, 120–122, 123–125, 146, 175
—, avoiding repeated items 6, 8, 97
—, avoiding structurally differentiated items 29, 31, 83
—, avoiding unnameable items 4, 9, 68, 92, 124, 128
—, contemporary positions 7–10, 24–31
—, criterion for, 120–123
—, for Goodman 24–31, 40–41, 82
—, medieval positions 4–6
—, vs. realism 6–10, 123–125, 128, 134, 135–136

Nominalistic model 142–143, *144–145*, 146–148, 151, 171, 181
—, system 120–122
Non-atomistic calculus of individuals 77–87
—, universe 73–78, *76*, 100
—, *see* also atom, atomistic
Null-item 35
—, *see* also empty set

Oblique context 111, 153, 175
Ontological commitment, according to Quine 102–110, 112, 114
—, criteria 102–125
—, denial 106, 116–119
—, doctrine 3, 124–125
—, implications of atomic sentences 112
—, implications of sentences 112–122
—, implications of theories 102–122
—, implications vs. commitments 108–110, 120
—, indirect implications 118–120
—, presuppositions, according to Cartwright 111–112, 114–116
Ontology, Leśniewski's system 7, 99
—, *see* also existence, nominalism
Operation *16*
—, closure under *16*
—, n-term 16
—, *see* also operation assignment, operation symbol
Operation assignment in classical models *20*, 21
—, in nominalistic models 144–146, 151
Operation symbol *17*, 150
—, nominalistic interpretation of, 136, 143–146, 151–153
—, realistic interpretation of, 20–21, 136
Order, problems of, 28–30, 81–99, 135–142, 192–197
—, provided by resemblance and matching 167
—, *see* also individual, sequential; pair, sequence, relation
Overlapping 45, 50, 150
—, in CB 150, 157

—, in CI I 50–62
—, in non-atomistic systems 77–79

Pair (ordered) *13*, *16*, 26, 82–83, 85–99, 137–139, 142
—, see also doubleton, sequence
Parsimony 6, 8, 128, 140
Part and whole, see part-whole relation
—, of (the symbol) 45–46, 78, 92, 150, 179
Part-Whole relation 3, *33*, 34–36, 38, 78, 84–85, 89, 90–91, *93*, 187
—, among concreta 187
—, among non-sets 178, 181
—, as a relation of predication 131–133, 186–188
—, of a model 56, 145, 151, 181
—, providing for order 29–30, 84–85, 89, 90–91, 93
—, qualitative *181*
Partial ordering *14*
Particular 6, 24, 185
—, see also individual
Physical object 24, 185
Platonism 10
—, see realism
Position 29, 84, 89–92, 185–186
Possible existence 119, 121
—, individuals 38–39, 99, 182
Power of the continuum *15*, 178
—, set *12*
Predicate *17*, 150
—, designating a bundle of qualities 152, 154–156, 186
—, designating an exemplar 124, 133–135, 136–137, 186–188, 191
—, designating a quality 124, 182
—, designating a sum 124, 131–133, 186
—, designating its extension 123, 127–128, 136
—, having multiple designation 5, 124, 128–130, 136–137
—, interpretation of 20–21, 89, 98, 113, 122–142, 152, 186–188, 190–197
—, pertaining to the *i*-th respect 190, 196
—, syncategorematic 107–108, 124, 125–127, 130, 136–138

Pre-systematic notions 1, 24–25, 27–31, 81–85, 92–93, 148–149, 155, 157, 166–167, 169, 185, 187
Primitive, see defined symbol
Problem of Universals: how to avoid them, see nominalism
—, see also universal
Product axiom 63, 69
—, of individuals 50, 62, 69, 100
—, relative *15*, 94
Property 175, 198
—, see quality
Protothetic 7

Quale (qualia) 29, 90–92, 124, 185–186
—, designatum of predicate 124, 131–133, 137
—, positional 29, 90–92, 185–186
Qualitative nearness or neighborhood 167, 171
—, part 177
—, part-whole relation *181*
Quality 149, 154–155, 175–177, 182
—, bundle or combination of, see bundle
—, calculus of (CQ) 175–184
—, designated by predicates 132–133, 160
—, individuated 149, 168–169
—, in the meta-theory 178
—, occurrent vs. dispositional 7, 149, 193
—, physical vs. phenomenal 149, 155
—, positive vs. negative 149, 155–156, 159, 175
—, simple vs. compound 149, 155–156, 168–169, 175–176, 182, 189
—, uniform 176, 181–182
—, uniform in a given respect 191
—, vs. property 175
Quantifier *17*
Quasi-quotation marks *18*
Quotation marks 18

R-least *33*
R-part *74*
Range of a relation *13*
—, of values, see universe
Realism (as opposed to nominalism)

INDEX OF SUBJECTS

123, 127–131, 134, 135–136, 142, 153
—, historically 4–10
—, *see* also nominalism
Reflexive relation *13*
Regularity, axiom of, 11, 100
Relation *13–14, 16,* 93, 96
—, as an individual 92–99, 138
—, as a sum of sequential individuals 98, 138, 192
—, closure under, *16*
—, generating, *see* generating relation
—, internal 140
—, interpretation of relation symbols 135–142, 145–146, 152–156, 192–197
—, IR-relation *93*
—, n-term relation *16*
—, of predication *127*, 128–135, 157, 186–188, 192
—, part-whole, *see* part-whole relation
—, symbols, *see* predicate
Relational system 115
Repeated item 8, 97
Representation theorem 36
Resemblance 2, 133–135, 148, 157, 166–175
—, as a relation of predication 133–135, 167, 188, 190
—, as indistinguishability 166–167, 169
—, calculus of (CR) 168–175
—, degree of 192, *193*
—, implying the existence of a respect 191
—, in a given respect 134–135, 188–189, *190*, 191
—, in the meta-theory 189
—, qualitative neighborhood 167, 171–172
—, vs. matching 157, 167, 170
Respect, commitment to respects of resemblance 191
—, *see* also resemblance, clan
Restriction of a relation to a set *14*
Rule, *see* inference rule

Satisfaction in models 22, *57*, 89, *95, 152–153, 172*
Schematic symbol 22, 150, 168, 186, 191
—, non-schematic symbol 22, 45, 50, 78, 95, 168, 190–191, 197
Semantical adequacy *23*, 44, 57–62, 65–67, 70–73, 80–81, 160–166, 170, 172–175, 184
—, notion of truth, *see* truth
—, yielding *22,* 157
Semantics, general remarks *20–23*
—, its relevance to nominalism 10, 28, 82, 86, 124–125
Sense 8, 129–130
Sequence *15–16,* 82–83
—, individual *97*
—, *see* also chain, individual, sequential, string
Sequential individual 28–30, 81–99, 137–139, 192
Set, endorsed in the meta-theory 11, 44, 178–179, 198
—, 'is a set' (Σ) *178*
—, proper and improper *12*
—, the empty set *12*
—, the set comprising exactly ... *13*
—, the set of all objects such that ... *12*
—, *see* also axiom, individual, set theory
Set theory 11–16, 99, 177–181
Similarity, *see* isomorphic, matching, resemblance
Simple quality, *see* quality
Singleton *12,* 24, 26
Soundness *23*, 44, 78, 101
—, of CB (bundles of qualities) 162–163
—, of CI I 60
—, of CI II 65–66
—, of CI III 70–71
—, of CI IV 80–81
—, of CQ (qualities) 184
—, of CR (resemblance) 172–173
—, of the calculus of individual relations 96
Spatial relations and individuals 9
Specifiable set 44, 50, 67–69, *68,* 70–72, 81, 196
Standard model, *see* model
—, of comparison, *see* exemplar
Strict inclusion *12*
—, partial ordering *14,* 27
String 97, 98

—, see also chain, sequence, sequential
Substance 4, 149
Substitution *18*, 156
Sum 24–25, *32*, 47, 50, 62, 100, 170, *180*, 189
—, and chain 196
—, and class 25
—, and concatenate 85–86
—, as supremum 32
—, axioms 47, 63, 69, 77, 169–170, 179
—, closure under 40–41, 50, 62–73, 91, 98
—, designated by a predicate 124, 131–133, 186
—, of scattered individuals 24–25, 40–41, 91, 98, 131–133, 138–139, 192
—, of sequential individuals 98, 138–139, 192
—, principle of summation 36, 39–42, 62–75, 91
Supremum *32*
—, see also sum
Symbol of the object-languages *17–18*
—, primitive vs. defined *20*
—, schematic vs. non-schematic *22*
Symmetric relation *14*
Syncategorematic 107–108, 109–110, 117, 125–127, 130, 137–138
System, abstractistic *187*
—, concretistic *187*
—, formal 17–18, *19*, 20–23, 109–110,
—, interpreted *23*, 102–104, 110
—, nominalistic 120–122
—, of individuals, see calculus, universe of individuals
—, relational 115

Temporal stages or parts 29, 84
Term *18*
—, denotationless 2, 46, 86, 142–147, 153
—, free vs. bound *18*
—, in calculi of individuals *45*
—, interpreted 145–146, 152–153
—, of a formal system *19*, 100
—, simple and compound 145–146, 153, 159
Theorem of a formal system *19*
Theory and ontological import 102-104

—, see also system
Thing 82
Togetherness 90–91, 185–186
Transitive relation *14*
Truth 22, 125–142, 153
—, condition 108, 125–142, 148, 152–157, 167, 186–188, 190, 193, 197
—, informative definition of 128–130, 133, 135
—, see also model, satisfaction, valid

Ultimate constituent, see atom
Uniform quality 176, 181–182, 191
Union, *12*, *13*
Unit 93, *94*, 95–101, *96*, 138
—, set, see singleton
Universal, as a non-individual designatum of a predicate 2, 123, 127–130
—, for Aristotle 4
—, individual 70
—, name 4, 6, 123
—, problem of universals (how to avoid them), see nominalism
Universe, atomistic 36–*42*, 49–50, 93–94, 144–145, 151, 171
—, empty 46, 56, 100, 143, 145
—, non-atomistic 73–78, *76*, 77–78
—, of actualized items 38–39
—, of discourse 20
—, of individual relations 94, *95*, 96
—, of nameable items 123, 128
—, see also existence, individual, model

Valid *22*, 57, 125, 126, 157
Value of a term 22, 144, 152
Variable *17*
—, assignment to, *20*, *56*, 108, 115, 145–*146*, 152, 171
—, ontological import of 102–108, 115, 117
Verification 129
Vocabulary of a term, formula, or system *19*
—, of CB (bundles of qualities) *150*
—, of CI I *50*
—, of CR (resemblance) *168*
—, of LGCI *45*

—, of the calculus of individual relations *95*

Well-founded relation 87

With, *see* togetherness
Whole, *see* part-whole relation, sum

Yielding *22, 57*

SYNTHESE LIBRARY

Monographs on Epistemology, Logic, Methodology,
Philosophy of Science, Sociology of Science and of Knowledge, and on the
Mathematical Methods of Social and Behavioral Sciences

Editors:

DONALD DAVIDSON (Princeton University)
JAAKKO HINTIKKA (University of Helsinki and Stanford University)
GABRIËL NUCHELMANS (University of Leyden)
WESLEY C. SALMON (Indiana University)

‡JAAKKO HINTIKKA and PATRICK SUPPES, *Information and Inference.* X + 336 pp. Dfl. 60.—

‡KAREL LAMBERT, *Philosophical Problems in Logic. Some Recent Developments.* 1970, VII+176 pp. Dfl. 38.—

P. V. TAVANEC (ed.), *Problems of the Logic of Scientific Knowledge.* 1969, XII+429 pp. Dfl. 95.—

‡ROBERT S. COHEN and RAYMOND J. SEEGER (eds.), *Boston Studies in the Philosophy of Science.* Volume VI: *Ernst Mach: Physicist and Philosopher.* 1970, VIII+295 pp. Dfl. 38.—

‡MARSHALL SWAIN (ed.), *Induction, Acceptance, and Rational Belief.* 1970, VII+232 pp. Dfl. 40.—

‡NICHOLAS RESCHER et al. (eds.), *Essays in Honor of Carl G. Hempel. A Tribute on the Occasion of his Sixty-Fifth Birthday.* 1969, VII + 272 pp. Dfl. 46.—

‡PATRICK SUPPES, *Studies in the Methodology and Foundations of Science. Selected Papers from 1951 to 1969.* 1969, XII + 473 pp. Dfl. 72.—

‡JAAKKO HINTIKKA, *Models for Modalities. Selected Essays.* 1969, IX + 220 pp. Dfl. 34.—

‡D. DAVIDSON and J. HINTIKKA: (eds.), *Words and Objections: Essays on the Work of W. V. Quine.* 1969, VIII + 366 pp. Dfl. 48.—

‡J. W. DAVIS, D. J. HOCKNEY, and W. K. WILSON (eds.), *Philosophical Logic.* 1969, VIII + 277 pp. Dfl. 45.—

‡ROBERT S. COHEN and MARX W. WARTOFSKY (eds.), *Boston Studies in the Philosophy of Science.* Volume V: *Proceedings of the Boston Colloquium for the Philosophy of Science 1966/1968.* 1969, VIII + 482 pp. Dfl. 58.—

‡ROBERT S. COHEN and MARX W. WARTOFSKY (eds.), *Boston Studies in the Philosophy of Science.* Volume IV: *Proceedings of the Boston Colloquium for the Philosophy of Science 1966/1968.* 1969, VIII + 537 pp. Dfl. 69.—

‡NICHOLAS RESCHER, *Topics in Philosophical Logic.* 1968, XIV + 347 pp. Dfl. 62.—

p.t.o.

‡GÜNTHER PATZIG, *Aristotle's Theory of the Syllogism. A Logical-Philological Study of Book A of the Prior Analytics*. 1968, XVII + 215 pp. Dfl. 45.—

‡C. D. BROAD, *Induction, Probability, and Causation. Selected Papers*. 1968, XI + 296 pp. Dfl. 48.—

‡ROBERT S. COHEN and MARX W. WARTOFSKY (eds.), *Boston Studies in the Philosophy of Science*. Volume III: *Proceedings of the Boston Colloquium for the Philosophy of Science 1964/1966*. 1967, XLIX + 489 pp. Dfl. 65.—

‡GUIDO KÜNG, *Ontology and the Logistic Analysis of Language. An Enquiry into the Contemporary Views on Universals*. 1967, XI + 210 pp. Dfl. 34.—

*EVERT W. BETH and JEAN PIAGET, *Mathematical Epistemology and Psychology*. 1966. XXII + 326 pp. Dfl. 54.—

*EVERT W. BETH, *Mathematical Thought. An Introduction to the Philosophy of Mathematics*. 1965, XII + 208 pp. Dfl. 30.—

‡PAUL LORENZEN, *Formal Logic*. 1965, VIII + 123 pp. Dfl. 18.75

‡GEORGES GURVITCH, *The Spectrum of Social Time*. 1964, XXVI + 152 pp. Dfl. 20.—

‡A. A. ZINOV'EV, *Philosophical Problems of Many-Valued Logic*. 1963, XIV + 155 pp. Dfl. 23.—

‡MARX W. WARTOFSKY (ed.), *Boston Studies in the Philosophy of Science*. Volume I: *Proceedings of the Boston Colloquium for the Philosophy of Science, 1961–1962*. 1963, VII + 212 pp. Dfl. 22.50

‡B. H. KAZEMIER and D. VUYSJE (eds.), *Logic and Language. Studies dedicated to Professor Rudolf Carnap on the Occasion of his Seventieth Birthday*. 1962, VI + 246 pp. Dfl. 24.50

*EVERT W. BETH, *Formal Methods. An Introduction to Symbolic Logic and to the Study of Effective Operations in Arithmetic and Logic*. 1962, XIV + 170 pp. Dfl. 23.50

*HANS FREUDENTHAL (ed.), *The Concept and the Role of the Model in Mathematics and Natural and Social Sciences. Proceedings of a Colloquium held at Utrecht, The Netherlands, January 1960*. 1961, VI + 194 pp. Dfl. 21.—

‡P. L. R. GUIRAUD, *Problèmes et méthodes de la statistique linguistique*. 1960, VI + 146 pp. Dfl. 15.75

*J. M. BOCHEŃSKI, *A Precis of Mathematical Logic*. 1959, X + 100 pp. Dfl. 15.75

Sole Distributors in the U.S.A. and Canada:
*GORDON & BREACH, INC., 150 Fifth Avenue, New York, N.Y. 10011
‡HUMANITIES PRESS, INC., 303 Park Avenue South, New York, N.Y. 10010